Sustainable Landscape Planning and Natural Resources Management

IEREK Interdisciplinary Series for Sustainable Development

Sustainable Landscape Planning and Natural Resources Management (SLNR) is a peer-reviewed book series that presents groundbreaking research through the intersection of two main areas that of which are 'Sustainable Landscape Planning' and 'Natural Resources Management'. The series initially covers a wide range of academic disciplines in sustainability, including landscape ecology, green urbanism, urban regeneration, land restoration and renaturation, energy management, and nature conservation. Moreover, it also incorporates the latest developments and research in engineering topics that contribute to more sustainable cities, such as Geosciences, infrastructure, civil engineering, geology, mining, water, and natural resources conservation and management.

By showcasing methods for discovering, managing, and employing natural resources, the series aims to provide approaches that enable the responsible use of these resources. Ultimately, the SLNR series is committed to advancing the United Nations' Sustainable Development Goals for 2030.

The wide range of topics covered in this series are primarily related to sustainable development, involving durable infrastructure projects, civil engineering, urban regeneration, the built, sustainability, climate change mitigation, energy management, geosciences, geology, and the future of cities, all of which depend on the availability of natural resources and effective wealth management. The series also features conceptual and empirical contributions from various interrelated fields of earth science, with an emphasis on the era of new discovery. It focuses on providing practical solutions to ensure sustainable development and provides new case studies that offer concrete examples of how to manage natural resources effectively.

The series encompasses the best research papers from various international conferences organized by IEREK conferences, also accepts proposals from other international conferences, including interdisciplinary contributions. It exhibits innovative approaches and highlights how they can best support both economic and sustainable development through the responsible use of natural resources and global, local, and individual activities for the welfare of all societies on Earth.

The SLNR series is a valuable source for experts in government, international organizations, consultancies, industry, business sectors, and academics. Published in cooperation with IEREK, the Springer SLNR series displays important new research in various sectors, delivering valuable insights and practical solutions to the complex challenges of sustainable development for the readers.

Mourad Amer · Mohamed El Kaftangui ·
Ibrahim El-Sayed Maarouf · Nauman Khalid ·
Sharul Sham Bin Dol
Editors

Territorial Landscapes Solutions for Sustainable Energy and Food Security

Springer

Editors
Mourad Amer
Architectural Engineering
Faculty of Engineering
Alexandria, Egypt

Ibrahim El-Sayed Maarouf
Alexandria, Egypt

Sharul Sham Bin Dol
Abu Dhabi, United Arab Emirates

Mohamed El Kaftangui
Abu Dhabi University
Abu Dhabi, United Arab Emirates

Nauman Khalid
Abu Dhabi University
Abu Dhabi, United Arab Emirates

ISSN 2948-1910 ISSN 2948-1929 (electronic)
Sustainable Landscape Planning and Natural Resources Management
ISBN 978-3-031-82418-0 ISBN 978-3-031-82419-7 (eBook)
https://doi.org/10.1007/978-3-031-82419-7

This Springer imprint is published by the registered company Springer Nature Switzerland AG
The registered company address is: Gewerbestrasse 11, 6330 Cham, Switzerland

Preface

The interconnected challenges of energy sustainability and food security have become paramount in the context of global environmental change. This volume, part of the *Sustainable Landscape and Natural Resources Management (SLNR)* series, addresses these critical issues by presenting innovative solutions that integrate territorial landscapes with sustainable practices. The challenging twenty-first-century goals need to be more concrete and focused on integrating environmental research with education and innovating solutions like resource management, climate management, providing effective resources for food security, the effective implementation of food policies, sustainable environment, an exemplary implementation of COP28 policies and suggestion might be few footsteps that will increase the worth of concepts like "Sustainable Energy" and "Food Security". Abu Dhabi University is leading in implementing the ideas of "Sustainable Solutions" by organizing the "International Conference on Advancing Sustainable Futures (ICASF 2023)," where more than 500 presentations highlight the Territorial landscape solutions for Sustainable Energy and Food Security.

This compilation of research highlights diverse methodologies and frameworks that facilitate the implementation of sustainable energy and food systems. The contributions are designed to provide actionable insights for researchers, policymakers, and practitioners, emphasizing the importance of interdisciplinary approaches in addressing complex sustainability challenges. This volume underscores the necessity of collaborative efforts to achieve sustainability goals by examining bio-based business models, agricultural management practices, soil science, and social perceptions of climate adaptation.

The editors believe this proceeding will be pivotal for all researchers, students, experts, and professionals in sustainability, engineering, environmental studies, and all allied disciplines. Moreover, the practical concepts in this proceedings will deepen the future understanding of managing sustainability and guide the policymakers in effective planning and execution of sustainable practices.

Alexandria, Egypt — Mourad Amer
Abu Dhabi, United Arab Emirates — Mohamed El Kaftangui
Alexandria, Egypt — Ibrahim El-Sayed Maarouf
Abu Dhabi, United Arab Emirates — Nauman Khalid
Abu Dhabi, United Arab Emirates — Sharul Sham Bin Dol

Introduction

Sustainable energy and food security are the most challenging subjects that affect society at its peak. The *Territorial Landscapes Solutions for Sustainable Energy and Food Security* volume presents innovative strategies and allied integrated approaches to comprehensively manage the concepts of sustainable energy and food security.

This volume is divided into three broad parts, and part I explores the *Bio-Based Business Models for Sustainability: Water and Climate Action*. This part highlights how integrated models and intelligent-based decision support systems can optimize energy management and, at the same time, promote effective and sustainable water and wastewater management practices. Moreover, this part reviews different case studies that show the significance of bringing different business models with Sustainable Development Goals (SDGs) regarding water resources and climate mitigation actions.

Part II of this volume highlights the *Multi-Disciplinary Approach: From Soil Amendments to Food Systems* and presents essential findings to achieve food security via managing soil health. This part presents different innovative solutions, like the utilization of halophyte biochar for improving millet-based cropping systems. Such mitigation solutions improve soil health and reduce environmental impacts that reduce soil fertility. Moreover, this part highlights the holistic framework of using food–energy–water systems to attain community-level sustainability.

Part III reviews *A Comprehensive Approach to Environmental Solutions* using a shift approach from social to environmental dimensions of sustainability. The public perception of climate change and management is an extensive topic that needs the integration of social science, environmental education, pastoral sociology, and community engagement. This part reviews the importance of societal engagement in achieving sustainability practices. The last chapter of this part highlights the role of politics in achieving constructive dialogue at international platforms like COP28 and the importance of diplomacy toward achieving sustainability.

These important contributions from renowned researchers provide a robust framework for better understanding and implementing diverse solutions for bridging gaps between energy sustainability and achieving food security. This volume will increase policymakers' basic knowledge of sustainability to effectively enforce energy sustainability policies and food security in territorial landscapes.

Contents

Bio-Based Business Models for Sustainability: Water and Climate Action

An Integrated Model of and for the Built Environment

John M. Kamara

Abstract

This paper proposes an extension of integration to the entire built environment. It is motivated by the need to appropriately respond to changes that impinge on and are caused by human civilization. An understanding of the consequences of change is crucial in the creation and management of a sustainable and resilient built environment, but no integrated model for this purpose exists. A proposal for an integrated model of the built environment, which draws from previous research on integration, is therefore presented. The potential benefits of this model and future research into its development and implementation are also outlined.

Keywords

Built environment · Integration · Model

1 Introduction

Integration generally refers to the joining-up of different people, processes, systems, applications, etc., to create a unified whole. The purpose is usually to facilitate more efficient and seamless collaborative working through a common interface (Gulledge, 2006) and can involve "*the continuous interdisciplinary sharing of data, knowledge, and goals among project participants*" (Fischer, 1989 in Betts et al., 1995, p. 4).

Within the Architecture, Engineering and Construction (AEC) industry, integration is seen as a key strategy to address the problems from fragmentation and has been a dominant theme for research and practice since the late 1980s. The anticipated benefits in improved and efficient processes, satisfied clients, and a sustainable and profitable industry (to name but a few) has led to research and advances into areas such as, product data modelling (Rezgui et al., 2011), information integration (Rempling et al., 2019), concurrent engineering (Anumba et al., 2007), collaborative working, integrated project delivery, process mapping (e.g., nD Modelling) (Aouad et al., 2006), and more latterly into Building Information Modelling (BIM), 4D construction, Digital Twins and so on (Georgiadou, 2019; Owen et al., 2013; Wen et al., 2021). Whilst the focus of such efforts has largely been on built assets (e.g., in integrating the data, information, knowledge, processes, tools, applications etc. that are involved in creation, management, and life cycle performance of assets), the quest has been both to maximize the benefits of such assets, create sustainable and resilient environments, and minimize (or eliminate) their negative impacts on the wider environment. Indeed, previous, and ongoing research on integration has led to many improvements in practice. For example, the use of BIM is becoming more widespread in the delivery of projects in the Global North, (Georgiadou, 2019) as is the use of other ICT-enabled technologies and applications (e.g., drones,). However, although the interrelationships between built assets and the wider society and environment are acknowledged and considered in their creation and management, integration efforts have not yet extended to include all these factors, although efforts have (and are being) made to address socio-technical factors in a research field that is mostly techno-centric (the application of ICTs to construction) (Li et al., 2019).

Given the benefits that integration (and its allied concepts such as collaboration) is bringing to the AEC and operations sectors (albeit against the acknowledged need for more inter-disciplinary and cross-sectoral research and higher levels of integration using advanced ICTs—Rempling et al., 2019), this paper seeks to extend this

J. M. Kamara (✉)
School of Architecture, Planning and Landscape, Newcastle University, Newcastle Upon Tyne, UK
e-mail: john.kamara@ncl.ac.uk

M. Amer et al. (eds.), *Territorial Landscapes Solutions for Sustainable Energy and Food Security*, Sustainable Landscape Planning and Natural Resources Management, https://doi.org/10.1007/978-3-031-82419-7_1

concept to the entire built environment. This is in recognition that a greater understanding of what constitutes the built environment and the interrelationships between the different systems therein, is vital to deal the intended and un-intended consequences of interventions made, and the pressures from climate and other changes that impinge on the built environment.

The structure of the paper of the rest of the paper is as follows: following a review of what constitutes the built environment and ways in which it is conceptualized, the issue of change is explored using a small-scale example from previous research into this area. Proposals for an integrated model of the built environment are then presented and described. The paper concludes with a discussion on the potential benefits of this model and future research into its development and implementation.

2 The Built Environment

The Built Environment (BE) is the "*man-made buildings and infrastructure stocks that constitute the physical, natural, economic, social and cultural capital*" of society (Hassler & Kohler, 2014). It is made up of different elements and systems such as buildings, infrastructure, and other artefacts (Fig. 1), with the implied emphasis on what is "man [human]-made" as opposed to the natural environment.

However, the agency of humans in the creation and use of the BE, from the natural environment, suggests that the BE can be conceived as also including the socio-technical system of the urban fabric and the socio-ecological relationship and interdependence between the built and unbuilt environment (Hassler & Kohler, 2014). Lamprecht (2016) observes that it "*…is not only the product of humanity but also the foundations of its development*" since it "*contains…*[created] *artefacts* [which provide] *material support for the production of social life in all its dimensions*" (p. 66).

The BE is characterized by relationships, between the elements within itself and between the 'man [human]-made' elements and natural environment. These relationships can also be "… between things and things, things and people, and people and people." (Rapoport, 1982, p. 178).

A key to the relationships associated with the BE is the interdependence between them. At a basic level, the characteristics of the natural environment (e.g., climate, resources) in any area largely determine the kind of BE produced and contribute to the development of people in that area. Similarly, the characteristics and culture of a society (e.g., demographics, values and lifestyle, and legal framework and systems), influence the production and consumption of the BE (Lamprecht (2016). There are, indeed, several studies that have explored the relationships between different facets of the BE at different scales–e.g., the influence BE on commuting trends (Susilo & Maat, 2007); residential property values (Diao & Ferreira, 2010); effect of indoor temperature on productivity (Porras-Salazar et al., 2021), and so on. Likewise, in acknowledgement of the interrelationships between the BE and the natural environment, people and social systems and cultures (Rempling et al., 2019), the consideration of the impacts of new developments (of built assets) on the environment, society, and the economy, is now required as part of the statutory approval processes such as planning and building control that operate in many countries.

Given the vital importance of the interconnections highlighted above, the BE is more than just the 'built human-made' artefacts and needs to be considered as "*a complex social–ecological system, where multiple-related metabolisms interact at different scales*" (Moffatt & Kohler, 2008, p. 248). A holistic understanding of how this complex web

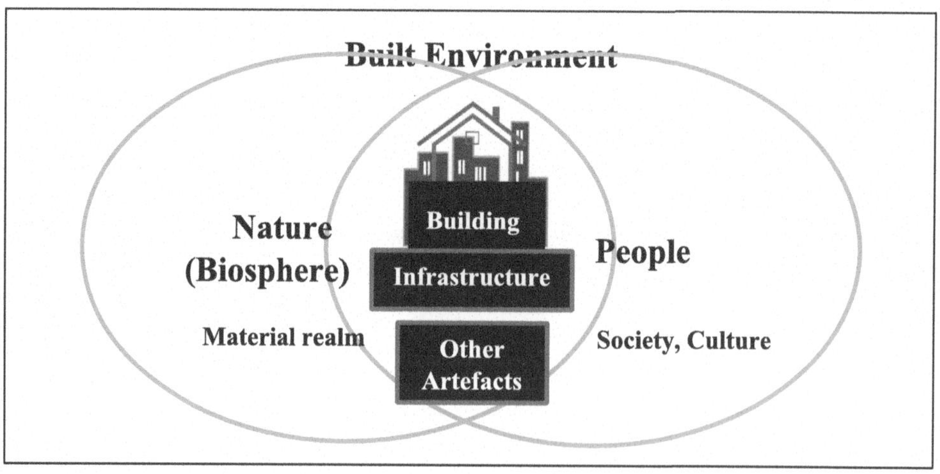

Fig. 1 Simplified model of the built environment. *Source* Adapted from a diagram by Fischer-Kowalski and Weisz (1999), cited in Moffatt and Kohler (2008)

of systems interact and the multiple ways in which each is affected by the other, is vital to a sustainable planet for all (Owen et al., 2013).

3 Need for an Integrated Model

The BE provides an important service and has a significant impact on the economic, social, and environmental sustainability of society–it positively contributes to quality of life but can also negatively affect the natural environment by consuming materials and energy in its creation, use, and operations (Rempling et al., 2019). The continued success (in terms of eliminating negative impacts) of the built environment depends on optimizing the socio-ecological relationship and interdependence between the built environment and nature, and on its resilience to adapt to changes from the climate and the environment to natural and man-made disasters as well as lifestyles, society, political, economic, or fundamental system changes, like, pandemics (e.g., COVID-19), or other inevitable but unforeseen changes.

This calls for more integration within and between the systems in the BE and its socio-ecological context. Owen et al. (2013) had proposed a Built Environment Information Fabric (BEIF), which will integrate building data, GIS (Geographical Information Systems), city modelling and other data sets to greatly enhance our "ability to comprehend a built form within the complete context of its placement in the environment" (p. 26) and *achieve a better balance between the natural and built environments*" (Rempling et al., 2019, p. 16).

The need for an integrated BE therefore builds is informed by these ideas, but it is also underscored by the need for more adaptability and resilience as key strategies to deal with change of all types and scales.

3.1 Change and Its Implications on the Built Environment

The Oxford English Dictionary defines change as "the action or process of making or becoming different" (OED, 2022). This is in relation to the properties of an object over time, either in its inherent properties or in its relationship to other objects (Worboys, 2005). In the context of the BE, this implies a change in the properties of (or within) built artefacts from minor renovations involving the repainting of a room, to major adaptations or refurbishments of entire buildings or neighborhoods (Kamara et al., 2020). It can also imply a change in the relationships and balance between different built artefacts (e.g., between buildings and infrastructure) and between built artefacts and nature and people/society (Fig. 1). For example, a new development of a building or land will constitute a change in the properties of a neighborhood and its relationship to nature. Similarly, a population, demographic or legislative change in society constitutes both a change in the inherent characteristics of society and in its relationship to built artefacts and nature.

Many factors can lead to change, but it is usually a response to a less than ideal or desirable state or quality and is affected to correct deficiencies (perceived or real) in performance (Gosling et al., 2013). In the built environment, triggers for change in any part of the system can be external to that system, and changes within a system can trigger changes elsewhere in the wider built/unbuilt environment–creating an interconnected and cyclical network of changes. In a study by Kamara et al. (2020), a conceptual model of change in relation to the adaptability of buildings (Fig. 2) was developed and used to investigate the details of changes to two buildings–a speculative office building (Building 1) and church building that had been converted into a heritage center (Building 2). Both buildings are owned by the same organization (a Local Government Authority—LGA). It is postulated in Fig. 2 that adaptations (changes) to the building fabric happen either directly from various triggers or when user adjustments to triggers are no longer adequate to maintain an acceptable level of performance.

Building 1 in this study (Kamara et al., 2020) was built in 2009 around the time of the world-wide economic recession that started in 2007/8. Although built by private developers, it was encouraged by the LGA who had designated the immediate area surrounding the building as an urban growth area. It was designed as a fully open plan to attract businesses that could occupy an entire floor (or floors). However, because of the economic conditions at the time, it was unoccupied for three years. Thereafter, the LGA decided to purchase the building and partition the spaces into smaller office units to attract smaller start-up companies, in a change to the business model on which the original design was based. Changes in the space plan led to other changes such as improvements to mechanical and electrical services, more toilet provision, soundproofing, and in the management of the building.

Building 2 had been in existence since around 1200AD and over time had undergone many changes and modifications, necessitated by fire damage (twice—in 1854 and 1979), expansion and change of use, and the need to bring it up to standard. Some of these changes were also influenced by the change in the urban dynamic of the area (e.g., relocation of residential buildings, and alterations and removal of tombstones from the graveyard, in the vicinity of the church building). It became a 'Listed' building in 1950 (i.e., added to the National Heritage List for England as a building of exceptional architectural or historic interest—Grade 1 listed

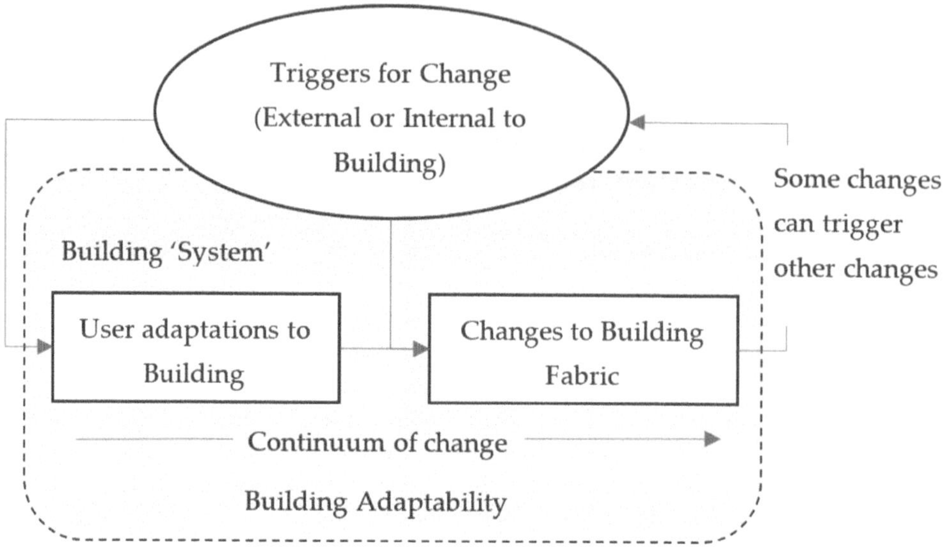

Fig. 2 Conceptual model of change and building and building adaptability. *Source* Kamara et al. (2020)

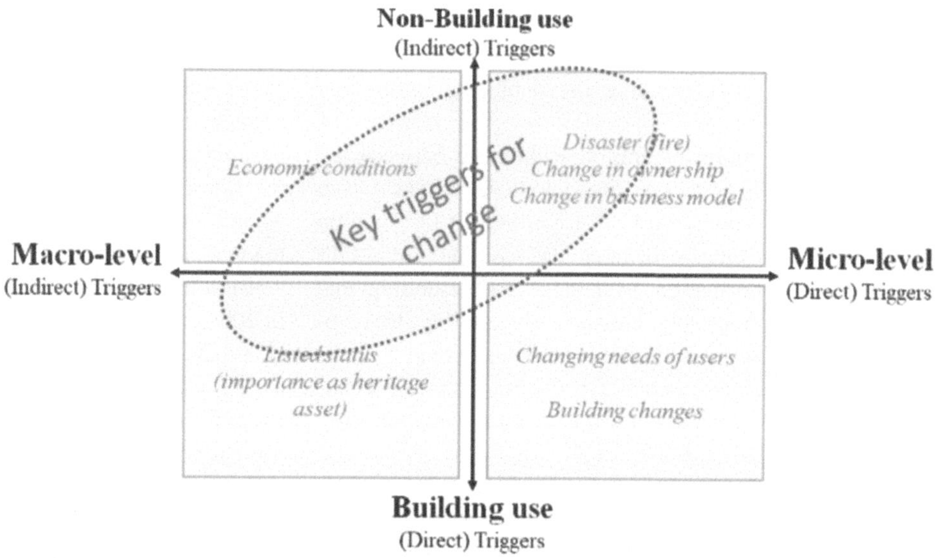

Fig. 3 Relationships between triggers for change in specific case studies reported in Kamara et al. (2020). *Source* Kamara et al. (2020)

building) and ceased to be a place of worship in 1979. The LGA purchased it in 2003 and subsequently converted it into a visitor and heritage center. The 'listed' status both facilitated the generation of funding for the refurbishment, but also severely constrained what could be done to the fabric. Following the economic recession in 2007/2008, the LGA decided to limit funding for the heritage activities and instead required the center to generate its own revenue. This necessitated further changes to accommodate different uses (e.g., for weddings and other events) (Kamara et al., 2020).

The findings from this study were that the triggers for changes to the fabric were mostly related to macro-level and non-building use factors such as economic conditions, fire disaster, change in ownership, and cultural (heritage status) factors (Fig. 3). A key conclusion was that "*a greater understanding of how non-building factors lead to changes in buildings will greatly enhance our ability to create more adaptable buildings*" (Kamara et al., 2020, p. 16). Extending this to the wider built environment, the likely implication is that non-built factors do affect built artefacts (and vice versa), and a greater understanding of the interrelationships between these will greatly enhance our ability to create a sustainable BE.

However, whilst the benefits of integrated models and ways of working have long been exploited within the AEC sector, these have mostly been at the micro scale of different facets of the built environment such as the construction project stage. Previous and ongoing considerations of life cycle assessment and more recent initiatives around the Circular Economy have sought to go beyond the short-term focus on projects (e.g., the idea of a BEIF by Owen et al., 2013). But extending these to the wider macro-scale of the built environment will provide a more holistic understanding of interrelationships between different systems within it.

3.2 Adaptability and Resilience

The inevitability of change is closely related to the current emphasis on adaptability and resilience of the built environment. **Adaptability** refers to the ability of buildings to easily accommodate the evolving demands of their context to maximize their value through life (Schmidt III et al., 2010). It is a key approach to deal with the problem of building redundancy, enhance the quality and performance and buildings (Douglas, 2006; Kronenburg, 2007), and to the sustainability of the built environment (Gosling et al., 2013; Langford et al., 2002; Webster, 2007; Wilkinson et al., 2009). The concept of adaptability has many dimensions, which relate to the different types of changes in or to buildings to suit changing needs (Arge, 2005; Schmidt III and Austin, 2016).

Similarly, the concept of **resilience**, which has evolved over time, has many meanings and has been applied to many fields and systems such as engineering, ecology, sociology, psychology, socio-technical, organisational, and social-ecological systems (Brand & Jax, 2007; Hassler & Kohler, 2014). At its core, it refers to the ability to respond, adapt and evolve to different forms of pressure (Hassler & Kohler, 2014), and within the built environment, it is considered as the ability to physically resist and rapidly recover from disruptive events (McAllister, 2013). The development of resilience in people, communities, cities, and infrastructure systems is central to the attainment of UN Sustainable Development Goals (SDGs) such as: building the resilience of the poor (SDG Target 1.5); building resilient infrastructure (SDG Goal 9); making human settlements resilient (SDG Goal 11) and strengthening the resilience and adaptive capacity to climate related hazards (SDG Target 13.1) (UN, 2015).

The reference to 'resilience' and 'adaptive capacity' in SDG Goal 9 and further evidence from socio-logical research indicate that there is a strong link between resilience and adaptability. For example, both Nelson et al. (2007) and Folke et al. (2010) suggest that adaptability is part of, and influences resilience. Fazey et al.'s (2007) definition of successful adaptability as the will to maintain resilience, further points to the idea that adaptive capacity does lead to resilience.

Since adaptability and resilience have to do with change (Pinder et al., 2017), the ability to assess and forecast change and its implication on different facets of the built environment will greatly enhance the creation and management of built artefacts. Given the interconnected nature of the facets and systems that comprise the built environment, and the interrelated nature of the changes that impinge on it, an integrated (i.e., joined-up) and holistic understanding of these interconnections and interrelationships can truly revolutionize the way the built environment is created and managed. However, previous research in both fields (e.g., Geraedts & Prins, 2015; Hassler & Kohler, 2014; Langford et al., 2002; Langston, 2012; Setunge et al., 2015) have not fully addressed the interconnected nature of the BE to benefit from the possibilities that this might achieve.

4 Integrated Model Development

Moffatt and Kohler (2008) observe that: "*An examination of dominant themes and methods for conceptualizing the built environment suggests that it is possible to build models of the built environment that allow changes in social, economic, and environmental conditions to be assessed. These models can be used to understand better the impact of different managerial and social policies at both the macro- and micro-levels.*" (Moffatt & Kohler, 2008, p. 265). Having outlined the need for an integrated model of the BE, the next set of questions that need to be considered include: *what* (e.g., what is it? what is it for? what form will it take? what is needed to make it work?), *how* (e.g., how will it be developed? how will it work?), and *who* (e.g., who is it for?) of the model.

Before addressing these questions (not necessarily in the order listed above), it is necessary to clarify how the term "integrated model" is used in this paper (the *'what is it?'* question mentioned above), since there are different aspects and scales to integration in construction and related sectors (Gulledge, 2006; Owen et al., 2013). 'Model' here refers to "*a representation of a system, which identifies the components of the system, its spatial and temporal bounds, the interactions among components, and the kind and range of dynamics that the interactions can experience*" (Pickett et al., 2004, p. 370). Building on the meaning of integration provided in the Introduction, 'integrated model' refers to a platform that will connect different systems within the built environment into a seamless whole.

4.1 Purpose and Tyne of Model

The purpose (*what is it for?*) of the Integrated Model being proposed is for analytical purposes. A model which will allow the investigation and analysis of the behavior of BE systems and the interrelationships between them (Guipponi et al., 2022), but primarily in relation to analyzing or appraising the implications of change (as described earlier in the paper). The vision is for the extension of the conceptual model in Fig. 2 into a model that will facilitate:

- A detailed understanding of the interrelationships between the triggers for change and the effects of such changes on other systems within the BE. For example, how do the characteristics of society influence the way land and buildings are used?
- An understanding of how triggers combine in any given situation to produce change, and the dynamics within the 'triggers for change'. For example, how do various elements or systems in the BE combine to influence other elements/systems? How might changes within one system influence others?
- Mapping and categorizing the responses to change by different systems. What actions have been taken within a system in response to triggers from within or outside that system?

The aim is for a *dynamic model* that will capture '*has-been*'/'*as-is*' realities, as well as the exploration or simulation of '*what if*' scenarios, to inform decision-making in relation to the creation and/or management of a sustainable, adaptable, and resilient built environment.

Regarding the type of model (*what form will it take?*), many options are possible for the kind of integration envisaged and what needs to be integrated. In their conceptualization of the built environment, Moffatt and Kohler (2008) describe different types of models that are applicable to the BE which include: flow-based models, resource-conservation-based models, top-down models, and bottom-up models (Moffatt & Kohler, 2008, p. 265). Other conceptions in the context of socio-ecological modelling for climate change adaptation (Guipponi et al., 2022) include biophysical process-based models, earth system models, dynamic global vegetation models, global circulation models, models of social systems, and integrated assessment models which "have the ambition to describe and simulate both human and the natural systems" (Guipponi et al., 2022, p. 4).

Given the desired functions of the model outlined above, and the fact that many models (and analytical tools) already exist for dealing with many of these interactions (Guipponi et al., 2022), ***a systems integration platform*** ("*a super-model of models*") (Gullledge, 2006) to interface with different (existing and developed) models of sub-systems within the BE, is likely to be the most viable option. However, it should be possible to interrogate this 'supermodel' from different perspectives or baselines, based on the BE system one is interested in—e.g., buildings, infrastructure, society, etc. Furthermore, as a *dynamic model*, it should have the facility for continuous data capture to maintain the relevance of '*as-is*' realities and analyses. An ***open-source*** platform, underpinned by an algorithm that builds the interrelationships between different systems or entities, that allows data entry and/or capture from multiple sources (e.g., sensors, linkups with other models or data sources, etc.), will therefore be necessary to facilitate the dynamic nature of the model.

4.2 Model Development (How?)

The process for achieving this could include:

1. The definition and development of an agreed taxonomy of the different systems and sub-systems that make up the built environment.
2. The identification and investigation of existing models and the required data sets and methodological underpinnings that currently exist for these systems. This might necessitate the development of new models either because of non-existence of any model for a defined system, or to have more data and/or methodological interoperability with other systems.
3. An investigation into, and the modelling of the characteristics and dynamics of BE systems, and the causes and effects of changes within them. This might also include a clearer understanding of what constitutes resilience and adaptability of BE systems.
4. The development and deployment of the integrated model to connect and allow the dynamic interactions between different systems.

Figure 4 shows a conceptual representation of the integrated model, with the entities of the BE (Fig. 1) depicted as systems. An "Nth System(s)" is included to acknowledge that the list of systems depicted is not a definitive list. The connections between systems and the integrated model reflect a

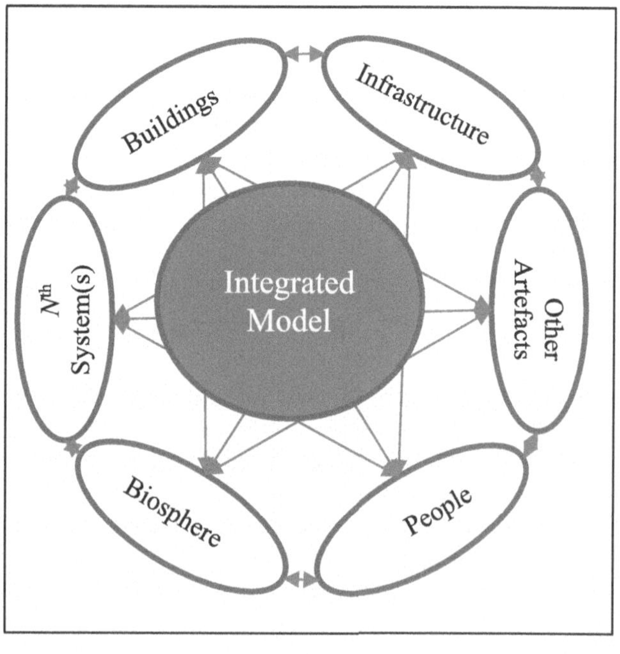

Fig. 4 A conceptual representation of the integrated model of the built environment interfacing with various BE systems. The inclusion of "Nth System(s)" acknowledges that the list of represented systems is not definitive

'hub-and-spokes' approach where the hub represents the integrating and processing platform that connects with different systems (or rather of different models within those systems).

But the lines between each system also acknowledge a 'many-to-many' relationship which reflects the reality that there may already be some level of interaction or integration between different systems (as outlined earlier).

Figure 4 is indeed a simplified representation as there are many scales and sub-systems within each system. For example, "buildings" incorporate the different types and scales of buildings, from individual units to neighborhoods, cities, and regions. Similarly, there are different types and scale of "infrastructure", and the "biosphere" incorporates many multiple scales including land, sea, and atmosphere. The research outlined in tasks 1–3 above (not necessarily implemented sequentially) will be instrumental in achieving a systematic nomenclature of systems and sub-systems that will enhance the development of the integrated model.

5 Conclusion

This paper has proposed an extension of integration to the entire built environment, which builds on previous and ongoing integration efforts within the AEC industry to achieve higher levels of integration. The basis for this and ideas on how this could be achieved have also been presented. It is acknowledged that the idea of integration beyond the design, construction and operation of built assets is not new. Previous researchers (i.e., Owen et al., 2013) have proposed similar ideas albeit within the context of a research roadmap.

This paper has sought to build on this by proposing ideas on how the development of an integrated module of the built environment could be achieved. It is based on a wider conceptualization of what is meant by the built environment, with a focus on change, adaptability and resilience and key areas that could underpin such a model. It is also underpinned by previous research into change factors and the adaptability of buildings, and the development of an integrated model will serve as an extension of this model.

It is further acknowledged that the achievement of the above is undoubtedly a very huge challenge and will require a multi- and trans-disciplinary approach, given that "*the scope of information sources envisaged would encompass councils and territorial authorities, transport authorities, utilities, land information resources, geotechnical resources, environmental resources, climate resources,* etc." (Owen et al., 2013, p. 13). It will require levels of collaboration that will equal (or surpass) the Human Genome Project.[1]

But the potential benefits of an integrated model of the built environment will be immense in out quest for a sustainable, adaptable, and resilient planet. This paper is therefore not only a proposition of an idea, but also a call for interested collaborators in this quest.

[1] https://www.genome.gov/human-genome-project.

References

Anumba, C. J., Kamara, J. M., & Cutting-Decelle, A.-F. (Eds.). (2007). *Concurrent engineering in construction projects*. Taylor and Francis.

Aouad, G., Lee, A., & Wu, S. (eds.), *Construction the future: nD modelling*. Taylor and Francis.

Arge, K. (2005). Adaptable office buildings: Theory and practice. *Facilities, 23*(3/4), 119–127.

Betts, M., Fischer, M. A., and Koskela, L. (1995). The purpose and definition of integration. In P. Brandon & M. Betts (Eds.), *Integrated construction information, E & FN SPON, London* (pp. 3–16)

Brand, F. S., & Jax, K. (2007). Focusing the meaning(s) of resilience: Resilience as a descriptive concept and a boundary object. *Ecology and Society, 12*(1), 23. Retrieved October 10, 2017, from http://www.ecologyandsociety.org/vol12/iss1/art23/.

Diao, M., & Ferreira, J. (2010). Residential property values and the built environment. *Journal of the Transportation Research Board, 2074*, 138–147. https://doi.org/10.3141/2174-18

Douglas, J. (2006). *Building adaptation* (2nd ed.). Abingdon.

Fazey, I. Fazey, J. A., Fischer, J., Sherren, K., Warren, J., Noss, R. F., & Dovers, S. R. (2007). Adaptive capacity and learning to learn as leverage for social-ecological resilience. *Frontiers in Ecology and the Environment, 5*(7), 375–380.

Fischer, M. (1989). A constructability expert system for the preliminary design of reinforced concrete structures. In *Proceedings of the 6th Conference on Computers in Civil Engineering* (pp. 60–66). ASCE.

Folke, C., Carpenter, S. R., Walker, B., Scheffer, M., Chapin, T., and Rockström, J. (2010). Resilience thinking: Integrating resilience, adaptability and transformability. *Ecology and Society, 15*(4), 20. Retrieved October 11, 2017, from http://www.ecologyandsociety.org/vol15/iss4/art20/.

Georgiadou, M. C. (2019). An overview of benefits and challenges of building information modelling (BIM) adoption in UK residential projects. *Construction Innovation, 19*(3), 298–320.

Geraedts, R. P., & Prins, M. (2015). The CE meter; an instrument to assess the circular economy capacity of buildings. In *Proceeding of the Going North for Sustainability Conference*, 23–25 Nov 2015, London South Bank University, UK.

Giupponi, C., Ausseil, A.-G., Balbi, S., Cian, F., Fekete, A., Gain, A. K., Essenfelder, A. H., Martínez-López, J., Mojtahed, V., Norf, C., Relvas, H., & Villa, F. (2022). Integrated modelling of social-ecological systems for climate change adaptation. *Socio-Environmental Systems Modelling, 3*, 18161. https://doi.org/10.18174/sesmo.18161

Gosling, J., Sassi, P., Naim, M., & Lark, R. (2013). Adaptable buildings: A systems approach. *Sustainable Cities & Society, 7*, 44–51.

Gulledge, T. (2006). What is integration? *Industrial Management & Data Systems, 106*(1), 5–20.

Hassler, U., & Kohler, N. (2014). Resilience in the built environment. *Building Research and Information, 42*(2), 119–129.

Kamara, J. M., Heidrich, O., Tafaro, V. E., Maltese, S., Dejaco, M. C., & Re Cecconi, F. (2020). Change factors and the adaptability of buildings. *Sustainability, 12*(16), 6585.

Kronenburg, R. (2007). *Flexible: Architecture that responds to change*. Laurence King
Lamprecht, M. (2016). The role of the built environment in human life: Selected issues. *European Spatial Research and Policy, 23*(2), 65–81.
Langford, D. A., MacLeod, I., Maver, T. W., & Dimitrijevic, B. (2002). Durability, adaptability and energy conservation (DAEC) assessment tool. *Internal Journal of Environmental Technology and Management, 2*(1/2/3), 142–159.
Langston, C. (2012). Validation of the adaptive reuse potential (ARP) model using iconCUR. *Facilities, 30*, 105–123.
Li, J., Greenwood, D., & Kassem, M. (2019). Blockchain in the construction sector: A socio-technical systems framework for the construction industry. In I. Mutis & T. Hartmann (Eds.), *Advances in Informatics and Computing in Civil and Construction Engineering*. Springer, Cham. https://doi.org/10.1007/978-3-030-00220-6_7.
McAllister, T. (2013). Developing guidelines and standards for disaster resilience of the built environment: A research needs assessment. In *NIST Technical Note 1795*. Retrieved October 11, 2017, from https://doi.org/10.6028/NIST.TN.1795.
Moffatt, S., & Kohler, N. (2008). Conceptualizing the built environment as a social–ecological system. *Building Research & Information, 36*(3), 248–268.https://doi.org/10.1080/09613210801928131.
Nelson, D. R., Adger, W. N., & Brown, K. (2007). Adaptation to environmental change: contributions of a resilience framework. *Annual Review of Environment and Resources, 32*, 395–419. Retrieved October 18, 2017, from http://www.annualreviews.org/doi/full/10.1146/annurev.energy.32.051807.090348.
OED. (2022). *Oxford English dictionary online*. Retrieved March 14, 2023, from https://www.oed.com.
Owen, R., Amor, R., Dickinson, J., Prins, M., & Kiviniemi, A. (2013). *Integrated design & delivery solutions (IDDS)*. CIB Publication 370. Retrieved February 28, 2023, from https://cibworld.org/roadmaps/.
Pickett, S. T. A., Cadenasso, M. L., & Grove, J. M. (2004). Resilient cities: Meaning, models, and metaphor for integrating the ecological, socio-economic, and planning realms. *Landscape and Urban Planning, 69*, 369–384.
Pinder, J. A., Schmidt, R., Austin, S. A., Gibb, A., & Saker, J. (2017). What is meant by adaptability in buildings? *Facilities, 35*, 2–20.
Porras-Salazar, J. A., Schiavon, S., Wargocki, P., Cheung, T., & Tham, K. W. (2021). Meta-analysis of 35 studies examining the effect of indoor temperature on office work performance. *Building and Environment, 203*, 108037.
Rapoport, A. (1982). *The meaning of the built environment—A nonverbal communication approach*. Sage Publications.
Rempling, R., Kurul, E., & Oti, A. H. (2019). *Research roadmap for information integration in construction*. CIB Publication 417. Retrieved February 28, 2023, from https://cibworld.org/roadmaps/.
Rezgui, Y., Boddy, S., Wetherill, M., & Cooper, G. (2011). Past, present and future of information and knowledge sharing in the construction industry: Towards semantic service-based e-construction? *Computer-Aided Design, 43*(5), 502–515.
Schmidt III, R., & Austin, S. (2016). *Adaptable architecture: Theory and practice*. Routledge.
Schmidt III, R., Eguchi, T., Austin, S., & Gibb, A. (2010). What is the meaning of adaptability in the building industry?. In Chica, Elguezabal, Meno & Amundarain (Eds.), *16th International Conference on Open and Sustainable Buildings*, CIB 104, Bilbao, Spain, 17–19 May (pp. 227–236)
Setunge, S., et al. (2015). Enhancing resilience of critical road infrastructure: Bridges, culverts and flood-ways under natural hazards. *Annual Project Report 2015–2016*
Susila, Y. O., & Maat, K. (2007). The influence of built environment to the trends in commuting journeys in the Netherlands. *Transportation, 34*, 589–609. https://doi.org/10.1007/s11116-007-9129-5
UN. (2015). Un sustainable development goals. Retrieved November 11, 2017, from https://sustainabledevelopment.un.org/.
Webster, M. D. (2007). Structural design for adaptability and deconstruction: A strategy for closing the materials loop and increasing building value. *New Horizons and Better Practices*, 1–6. Retrieved October 30, 2014, from http://ascelibrary.org/doi/abs/10.1061/40946(248)27.
Wen, Q.-J., Ren, Z.-J., Lu, H., & Wu, J.-F. (2021). The progress and trend in BIM research: A bibliometrics-based visualization analysis. *Automation in Construction, 124*, 103558.
Wilkinson, S. J., James, K., & Reed, R. (2009). Using building adaptation to deliver sustainability in Australia. *Structural Survey, 27*(1), 46–61.
Worboys, M. F. (2005). Modelling changes and events in dynamic spatial systems with reference to socio-economic units. In A. U. Frank, J. Raper, & J.-P. Cheylan (Eds.), *Life and motion of socio-economic units*: *GISDATA* (vol. 8, pp. 115–123). Taylor and Francis.

An Intelligent Decision Support System for Sustainable Energy Management

Doğa Aksu, Melih Akgün, Ufuk Cebeci and Nedim Kebabci

Abstract

By creating a data-driven, intelligent decision support system (IDSS), this project focuses on improving energy management for businesses by utilizing forecasting techniques and insights derived from monitoring energy data. Electricity consumption predictions are made by using a variety of inputs including historical consumption data of manufacturing companies. This will reduce the negative effects of uncertainty, resource waste, and potential power outages. This creates a more sustainable environment by preventing excessive energy consumption. Various algorithms are tested using Python, focusing on time series using machine learning (ML) techniques. Then, the results are compared to determine the model that predicts the data best. The future electricity consumption can be estimated using the selected model in IDSS. Later, companies can perform budget and invoice management using this estimated data. For application, by gathering data from Hudop Technology Inc.'s database and performing necessary preprocessing steps, the goal is to select the best model having high accuracy and low error for every scenario and use it in IDSS for electricity consumption forecasts of manufacturing companies.

Keywords

Energy management · Decision support system · Forecasting · Machine learning · Sustainability

1 Introduction

Energy sector consists of electricity, oil and natural gas industries. The aim of this study is to predict electricity consumption using ML methods for enabling companies to limit their usage and to plan in advance according to the predictions made. In this way, a more sustainable environment can be achieved.

Main problems of the energy market are inefficiencies in energy production and usage, uncertainty in the amount of energy consumption, production, and accordingly in prices. Energy consumption has a significant impact on other factors and there is a gap in building energy usage forecast in the literature (Forootan et al., 2022). Therefore, a data driven intelligent decision support system is designed which creates electricity consumption forecasts for companies to provide them better energy management solutions.

This project is carried out with Hudop Technology Inc. which provides an energy consumption tracking platform Powerhud for companies using IoT. In this way, an infrastructure suitable for Digital Twin will be created. Energy management insights can be obtained through monitoring energy data, and the predictions can be used not only to create better solutions for today but also to create new solutions for the future.

2 Intelligent Decision Support System Design

2.1 Literature Review

Time series, ML, and DL methods are used for electricity consumption forecasts in literature. Exponential smoothing is one of the best prediction techniques if a time series

D. Aksu (✉) · M. Akgün · U. Cebeci
Istanbul Technical University, Maslak, Turkey
e-mail: aksud18@itu.edu.tr

M. Akgün
e-mail: akgunme19@itu.edu.tr

U. Cebeci
e-mail: cebeciu@itu.edu.tr

N. Kebabci
Hudop Technology Inc., Kirazpinar, Turkey
e-mail: nedim@hudop.com

M. Amer et al. (eds.), *Territorial Landscapes Solutions for Sustainable Energy and Food Security*,
Sustainable Landscape Planning and Natural Resources Management, https://doi.org/10.1007/978-3-031-82419-7_2

has an altering trend. For time series having a linear trend and seasonal variance Holt-Winters and ARIMA models are appropriate. ML and DL methods can be used for every scenario, with each having different performances under different scenarios (Forootan et al., 2022).

Amasyali et al. examined studies on energy consumption prediction for buildings using the majority of ML and DL algorithms, focusing on commercial and educational structures. Short-term energy usage forecasting received more attention compared to long-term predictions. The findings suggest that ML models are effective for this purpose, but each has its own strengths and weaknesses. Thus, it is important to select the appropriate model for a specific application (Amasyali & El-Gohary, 2018). Walker et al. utilized data from 47 commercial buildings collected over two years to forecast hourly electricity usage. ML techniques such as BT, RF, SVM, and ANN were employed. Results showed that the RF model outperformed others in terms of prediction accuracy (Walker et al., 2020). Diego F. Lizondo et al. employed RF to anticipate energy consumption in the province of Tucumán, Argentina, with a high degree of success. They were able to improve the temporary variable by up to 10.56% by dividing it into the variables like day, day of the week, month, and year separately. This is applied in preprocessing for a more accurate prediction (Lizondo et al., 2015).

González-Briones et al. compared LR, kNN, RF, SVR, and DT models using a one-year dataset of daily power consumption from a Spanish shoe business. The models were trained with optimized parameters and normalized values. The LR and SVR models yielded the best results, achieving an accuracy of 85.7% (González-Briones et al., 2019). Liu et al. identified patterns in public buildings' energy usage and evaluated how well the SVM algorithm predicts it. The outcomes demonstrate that this method is capable of determining if energy usage is normal or abnormal with acceptable accuracy and error (Liu et al., 2020). Fuzzy, CBR, SVM, MA, ES, Grey, ANN, HM, and ARIMA methods were reviewed by Deb et al. for building energy consumption, considering qualitative and quantitative comparisons for each. One notable finding was that methods such as ANN and ARIMA perform well when combined with optimization methods (Deb et al., 2017).

By considering the building types, energy types, and time horizon; Fathi et al. reviewed energy performance prediction studies of urban structures. The findings indicate that ANN and SVR are utilized most frequently (Fathi et al., 2020). To forecast and examine energy supply and demand, Grimaldo et al. integrated the kNN algorithm with visual analytics. In order to detect consumption and production patterns, the user can assess several forecasting alternatives and refer them to the inputs, thanks to the results' acceptable accuracy (Grimaldo & Novak, 2020). Jozi et al. propose a Java-based decision support system that enables users to forecast energy usage in buildings using five prediction techniques. A case study was conducted using hourly consumption data from a building in Portugal and local temperature as input. SVM yielded the best result with a 4.35% MAPE error when the third strategy was implemented. The DSS allows users to input data, choose desired methods for forecasting and compare the results using MAPE (Jozi et al., 2019).

3 Methodology

The IDSS is designed to be integrated with the energy usage tracking platform of Hudop Technology Inc. as they provide the database and user interface. The company database consists of internal and external data. Internal data includes stakeholder company characteristics as well as sensor, analyzer, and meter data of electrical appliances used in each of them, while external data consist of weather conditions and tariffs. Company data includes the relevant characteristics such as the working hours, machine status, fault history and electricity usage data. In the Powerhud platform of Hudop Technology Inc., the usage data and recommendations can be visualized in dashboards. Recommendations can be created on necessary issues of energy management such as bill and demand management. Companies can use these in many decisions from allocating their budgets and planning their operations (Fig. 1).

IDSS is developed using the best model possible for predicting future electricity usage. For choosing the best model to be used in IDSS, different algorithms are used and compared in Python using relevant libraries like scikit-learn. ML methods are preferred based on the literature review, seeing that they are frequently utilized for predicting electricity usage. First, data is gathered from the database of Hudop Technology Inc. linked according to dates. Gathered data are prepared by necessary steps such as data cleaning, conversion and feature selection. These preprocessing steps depend on the inputs. After treating the missing values, categorical data are converted into numerical values named dummy variables. Feature selection is applied to determine the most useful input variables, which have the strongest influence on the output variable, hence the most informative values. As a result, models can be trained faster and afterwards easily generalized to real life, because more features take more time to learn and result in overfitting, which is the problem of memorizing the existing data without extracting the pattern in it. Models are developed by selecting optimal hyperparameters and training them accordingly. This is done after splitting the data into train, validation, and test sets. Then, RMSE, MAPE, and R2 values are calculated to compare and choose a model according to their

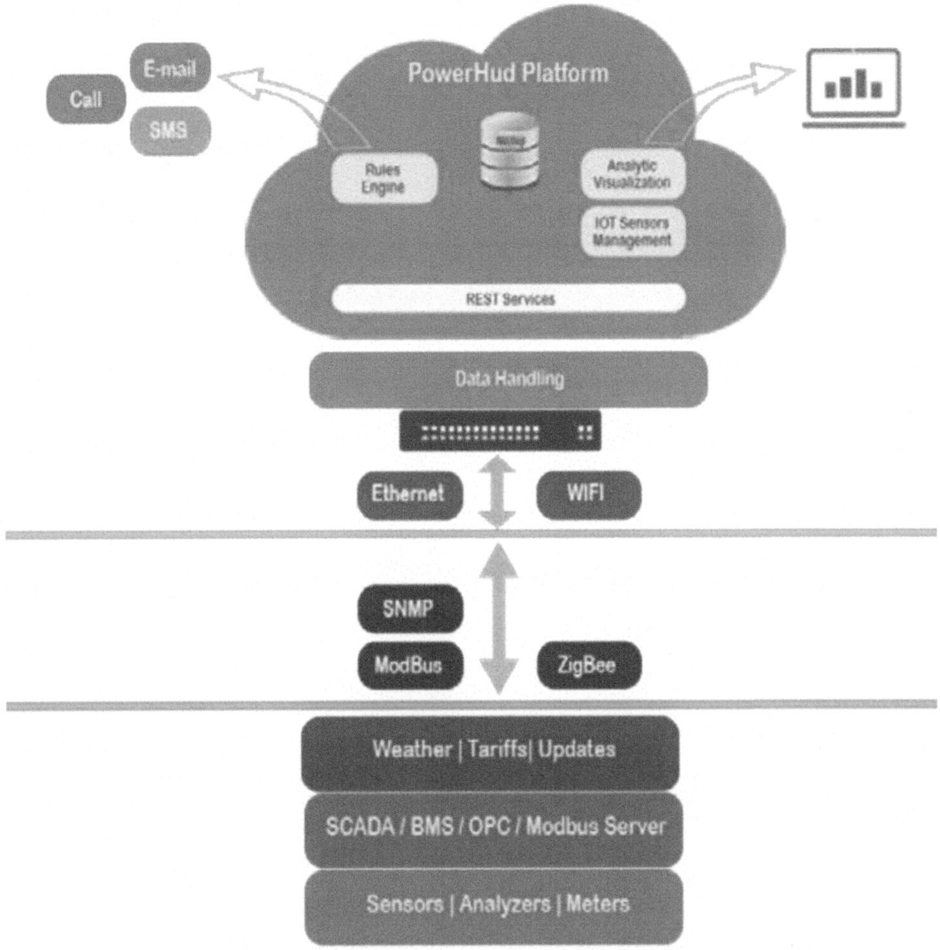

Fig. 1 Powerhud energy management software

performance in the validation set, which results in the most generalizable algorithm. Models having the highest R2 and lowest errors would be the most proper choice to use in future predictions, depending on the input combinations. IDSS is developed using the chosen models and preprocessing steps, which will be updated whenever new features are defined. With the chosen model, forecasts can be created and analyses can be performed using these estimations (Fig. 2).

The IDSS will carry out the necessary steps to create forecasts, as can be seen in Fig. 3, whenever new inputs are provided. Necessary features and their combinations can be determined by an inference engine considering the input characteristics of each company. The data is then converted by preprocessing into a format that can be used in estimations considering selected features by inference engine. Using these, electricity usage forecasts are performed with the chosen model and stored in the database. Predictions can then be fed into the inference engine to create recommendations. For sustainability, some suggestions can be given such as: "Prepare Plan B for lack of electricity because of wars near your country or/and some pandemic disasters such as Covid-19". Afterwards, the forecasting results and recommendations will be integrated with the reports using graphs such as usage trends so that users can analyze them. According to user feedback, the forecasts and recommendations can be updated by changing the relevant steps if results are not satisfactory.

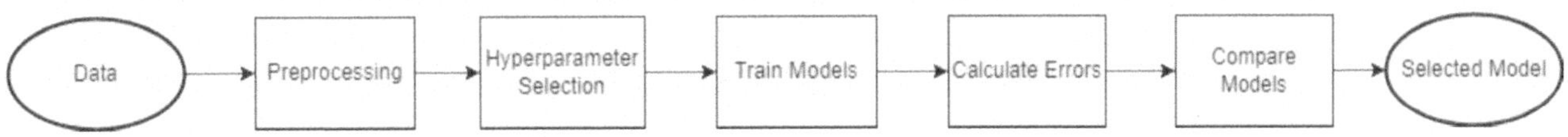

Fig. 2 Steps of developing the model

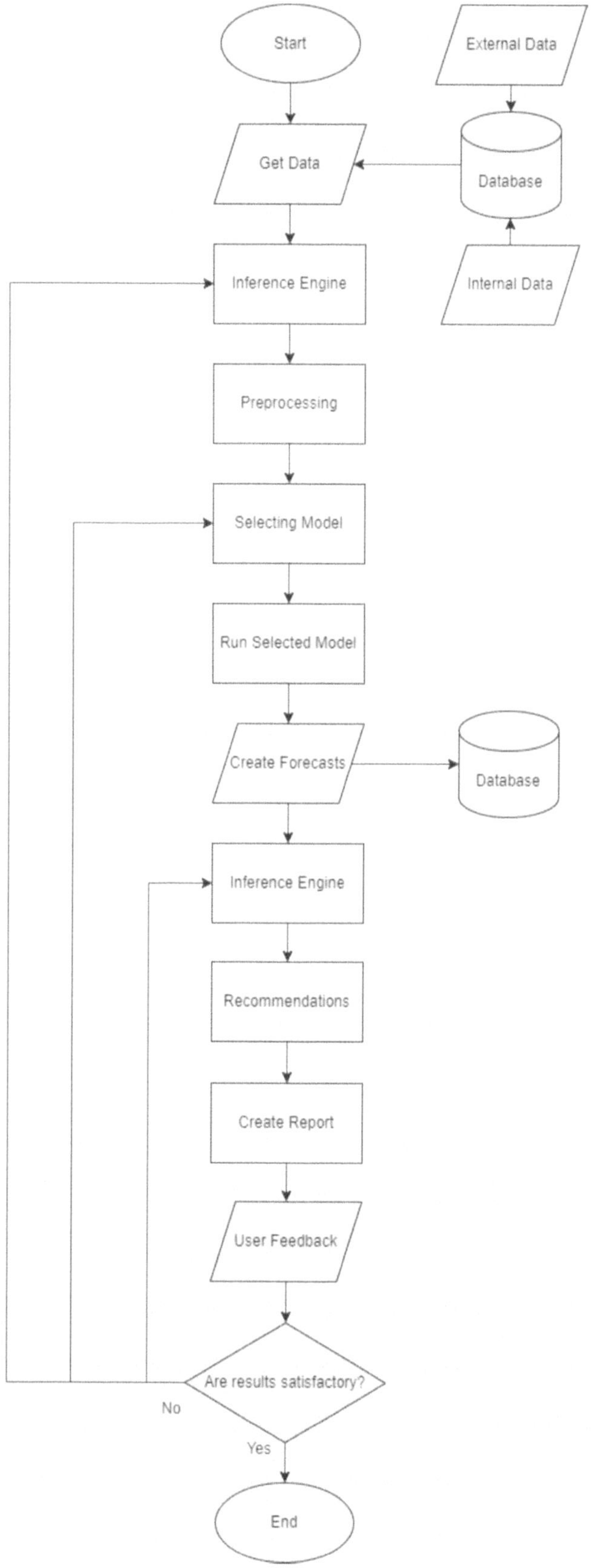

Fig. 3 IDSS

4 Application

For application, steps in the IDSS flowchart are performed. First of all, electricity usage data of a polymer manufacturing company was received from Hudop Technology Inc., which has ambitious sustainability targets for 2030, focusing on reducing greenhouse gas emissions and reducing landfill waste intensity. Among their data, values of 2022 are used for training and 2023 for testing. In the preprocessing, the data is made suitable for analysis with these steps:

- The retrieved dataset had logs of "Timestamp" and "Total Consumed Active Energy". In order to use this data, the "Timestamp" column is divided into 6 different columns: year, month, day, hour, minute and second. The difference of electricity usage values are taken to determine the amount used in the time intervals to reach the kW values.
- A "tdiff" column is created by taking the time difference of the data in the "Timestamp" column with the help of the "mod" function and converted into seconds.
- A "weekday" column is created by assigning numbers from 1 to 7 to the days of the week. Thanks to this new feature, the models achieved higher accuracy by finding the correlation between the same days of the week.
- A "holiday" column was created to increase accuracy of the analysis. To the days such as public, religious, national holidays and weekends are assigned 1 and the remaining days are assigned 0 as dummy variables.
- Minute and second information is removed by grouping data hourly, which creates more informative values. Grouping the data daily was not preferred since it would reduce the number of data drastically.
- At the end, 3 different scenarios are created. First, the input is given row by row without kW values. Secondly, previous kW values are grouped in 12 and included in the input. At last, outputs are also grouped in 12 to create prediction for multiple hours (Fig. 4).

Using these data, different ML models are developed for each scenario. After preprocessing, datasets are splitted into train, validation and test sets. Then, using the Grid Search algorithm, hyperparameter selection is made for 6 ML models considering different combinations of possible parameters using a 5-fold cross validation method, which splits the data into 5 parts and calculates scores for each part to get a more trustable score by taking their average. The chosen parameters are used to train each model. Training results are used to assess whether or not models are underfitted or overfitted and update the parameter pool accordingly, while validation results are used to compare models (Figs. 5, 6 and 7).

According to 20-fold cross validation results of trained models, Random Forest algorithm performed the best in 1st

	kw	tdiff	day	month	year	hour	weekday	holiday
0	2	3031	1	1	2022	0	6	1
1	3	3661	1	1	2022	1	6	1
2	2	3481	1	1	2022	2	6	1
3	3	3631	1	1	2022	3	6	1
4	2	3632	1	1	2022	4	6	1
...	...	...	...	...	...	...	...	...
8638	3	3724	31	12	2022	19	6	1
8639	2	3662	31	12	2022	20	6	1
8640	2	3630	31	12	2022	21	6	1
8641	3	3421	31	12	2022	22	6	1
8642	2	3722	31	12	2022	23	6	1

Fig. 4 Dataset

and 3rd scenario and Gradient Boosting algorithm in 2nd scenario. Best models are tested in 2023 data, in which the 2nd scenario performed best with 94.33 % R2 score, which uses previous 12 hours' data to predict the next hour. While the 1st scenario can predict further into the future, its performance is relatively lower with 60.52 % R2 score. As further research, more features can be added such as weather conditions and data from various companies such as Goldenline Jewellery manufacturing company can be used to improve results (Fig. 8).

5 Conclusion

In this study, models are created to be used in IDSS to forecast future electricity usage. Based on a thorough literature analysis, different algorithms are created and compared in

Fig. 5 Results of 1st scenario

	Train R2	Validation R2	Train MSE	Validation MSE
Random Forest	0.968972	0.894515	1883.760503	13666.398532
Linear Regression	0.572974	0.426257	25925.282126	37811.197637
Decision Tree	0.915248	0.790011	5145.398358	19639.420404
Gradient Boosting	0.956676	0.848802	2630.278452	15497.460626
k Nearest Neighbors	0.611518	0.207167	23585.211663	42228.232771
Support Vector Machines	0.102055	-0.075471	54515.330579	63618.760232

Fig. 6 Results of 2nd scenario

	Train R2	Validation R2	Train MSE	Validation MSE
Random Forest	0.953611	0.897389	2697.655646	19446.849546
Linear Regression	0.792850	0.356966	12046.496840	46287.514790
Decision Tree	0.988621	0.868872	661.701411	20990.895179
Gradient Boosting	0.976616	0.918895	1359.856591	14301.651093
k Nearest Neighbors	0.869275	0.899965	7602.107275	14915.250381
Support Vector Machines	0.524099	0.730734	27675.341739	27937.940757

Fig. 7 Results of 3rd scenario

	Train R2	Validation R2	Train MSE	Validation MSE
Random Forest	0.938651	0.786826	3645.499824	25802.633142
Linear Regression	0.514687	0.637658	28863.665359	32628.772873
Decision Tree	0.801581	0.459835	11817.422311	39121.879391
Gradient Boosting	0.729328	0.770239	16115.467108	26278.106582
k Nearest Neighbors	0.646011	0.546311	21060.164889	35428.398054
Support Vector Machines	0.370255	0.506640	37412.051435	37761.564032

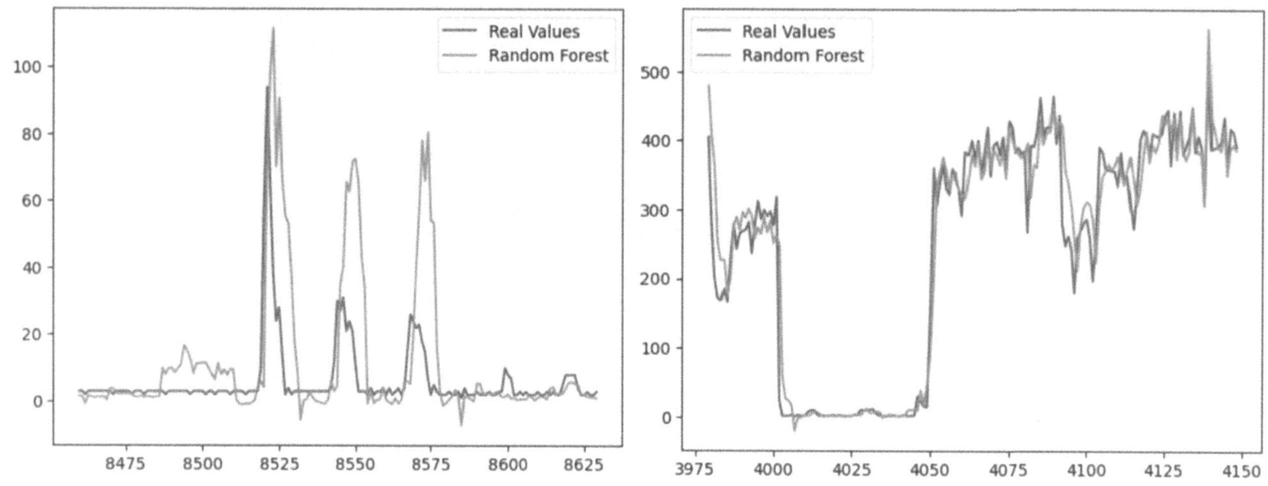

Fig. 8 Comparison of train and test values with predictions

Python, using ML techniques. The model with the highest R2 and lowest MSE is chosen to be used in IDSS, which will be integrated with the energy consumption tracking platform of Hudop Technology Inc. for interpretation. In this way, infrastructure suitable for Digital Twin will be created and companies can get the best results possible as this IDSS will choose the best model appropriate to their preferred input combinations and apply necessary preprocessing steps. They can use these forecasts in their strategic and operational decisions by analyzing trends and anomalies. In this way, they can lower their costs and achieve a more sustainable system by implementing necessary changes.

This project addresses the feasibility of energy forecasting in terms of its management advantages, environmental, economic, and social impacts, and ethical concerns. Energy consumption forecasting provides many benefits for a sustainable society which can be measured according to KPIs. Chief among these are CO_2 emissions reduction, energy consumption reduction, waste reduction, and environmental benefits. By estimating energy usage, unnecessary energy consumption is detected and waste is prevented. Thus, CO_2 emissions are reduced. CO_2 emission reduction also decreases the effects of global climate change, improves public health, strengthens the global economy, and protects biodiversity. Reducing carbon emissions helps provide cleaner air, water, and food for future generations.

In the energy sector, the ability of machine learning algorithms to predict future energy demands through the analysis of large amounts of data is emphasized. While energy forecasts facilitate resource management by improving energy efficiency, they also provide better understanding by revealing patterns and trends in energy usage. This also increases energy security and contributes to environmental and economic sustainability. However, ethical concerns such as data security, bias, and transparency of decision-making processes must also be considered so that energy forecasts can be achieved fairly, reliably, and sustainably.

References

Amasyali, K., & El-Gohary, N. M. (2018). A review of data-driven building energy consumption prediction studies. *Renewable and Sustainable Energy Reviews, 81*, 1192–1205.

Deb, C., Zhang, F., Yang, J., Lee, S. E., & Shah, K. W. (2017). A review on time series forecasting techniques for building energy consumption. *Renewable and Sustainable Energy Reviews, 74*, 902–924.

Fathi, S., Srinivasan, R., Fenner, A., & Fathi, S. (2020). Machine learning applications in urban building energy performance forecasting: A systematic review. *Renewable and Sustainable Energy Reviews, 133*, 110287.

Forootan, M. M., Larki, I., Zahedi, R., & Ahmadi, A. (2022). Machine learning and deep learning in energy systems: A review. *Sustainability, 14*(8), 4832.

González-Briones, A., Hernandez, G., Corchado, J. M., Omatu, S., & Mohamad, M. S. (2019). Machine learning models for electricity consumption forecasting: A review. In *2019 2nd International Conference on Computer Applications & Information Security (ICCAIS)* (pp. 1–6). IEEE.

Grimaldo, A. I., & Novak, J. (2020). Combining machine learning with visual analytics for explainable forecasting of energy demand in prosumer scenarios. *Procedia Computer Science, 175*, 525–532.

Jozi, A., Pinto, T., Praça, I., & Vale, Z. (2019). Decision support application for energy consumption forecasting. *Applied Sciences, 9*(4), 699.

Liu, Y., Chen, H., Zhang, L., Wu, X., & Wang, X. J. (2020). Energy consumption prediction and diagnosis of public buildings based on support vector machine learning: A case study in China. *Journal of Cleaner Production, 272*, 122542.

Lizondo, D. F., Jimenez, V. A., Villacís Postigo, F., Will, A. L., & Rodriguez, S. A. (2015). Análisis de variables temporales para la predicción del consumo eléctrico.

Walker, S., Khan, W., Katic, K., Maassen, W., & Zeiler, W. (2020). Accuracy of different machine learning algorithms and added-value of predicting aggregated-level energy performance of commercial buildings. *Energy and Buildings, 209*, 109705.

Implementation of SDGs on Water Resources and Climate Action: Sustainable Water and Wastewater Management Based on Swampland Adaptation at UIN Raden Fatah Palembang

Nyayu Khodijah, Ledis Heru Saryono Putro, Ris'an Rusli, Syahril Jamil, Ria Agustina and Siti Soleha

Abstract

Universities have an essential role in achieving the SDGs. Management and control of water and wastewater resources based on swampland adaptation is the 6th SDGs, Clean Water and Sanitation, and the 13th SDGs, Climate Action. This paper examines water resources management based on wetland adaptation in the urban campus area of UIN Raden Fatah Palembang (UINRFP). The research used primary and secondary data and conducted water and wastewater laboratory tests and data analysis. Rainwater harvesting is carried out to prevent flooding on the campus, as well as to provide water reserves for life. The volume of rainwater harvested and stored in campus water canals is 30,424 m^3/year, a potential economic water reserve and renewable energy source. Construction techniques with precast concrete pit foundations type squares pile a 32–40 m 1461 in all campus buildings and utilities. The construction technique is a solution for urban swampland adaptation, estimated to withstand water infiltration at a slow rate of 3357.5 m^3. The Sewage Treatment Plant (STP), with a capacity of 506 m^3, was built to treat domestic wastewater. Domestic wastewater is collected in 25 ground tanks (capacity 2–21 m^3) and channeled to STP for further processing. After meeting quality standards, it is stored in a retention pond to become campus reserve water. The quality of water resources is tested in periodic laboratories; all test results meet water or wastewater quality standards. Water resources and their management on the UINRFP campus as a potential water reserve of 75,171 m^3 for micro-hydro power generation, raw-clean water, drinking water production for campus needs, and even business unit development. Through the above efforts for efficiency, conservation, and usability of water resources, the university contributes to reducing the impact of climate change.

Keywords

Rainwater harvesting · Urban swampland adaptation · Water canals · Retention pond · Sewage treatment plant · Micro-hydro power

1 Introduction

Sustainable development goals (SDGs) are a development system that ensures increased economic prosperity, sustainable social life, environmental quality, development that ensures justice, and the implementation of a governance

N. Khodijah
Department of Islamic Religious Education, Faculty of Tarbiyah and Teacher Training, Universitas Islam Negeri Raden Fatah Palembang, Palembang, Indonesia
e-mail: nyayukhodijah_uin@radenfatah.ac.id

L. H. S. Putro (✉) · S. Soleha
Department of Biology, Faculty of Science and Technology, Universitas Islam Negeri Raden Fatah Palembang, Palembang, Indonesia
e-mail: lherusp@radenfatah.ac.id

S. Soleha
e-mail: sitisoleha@radenfatah.ac.id

R. Rusli
Master Program Al-Qur'an and Tafsir Science, Faculty of Ushuluddin and Islamic Thought, Universitas Islam Negeri Raden Fatah Palembang, Palembang, Indonesia
e-mail: risanrusli_uin@radenfatah.ac.id

S. Jamil
Department of Islamic Family Law, Faculty of Sharia and Law, Universitas Islam Negeri Raden Fatah Palembang, Palembang, Indonesia
e-mail: syahriljamil_uin@radenfatah.ac.id

R. Agustina
Department of Zakat and Waqf Management, Faculty of Islamic Economics and Business, Universitas Islam Negeri Raden Fatah Palembang, Palembang, Indonesia
e-mail: riaagustina_uin@radenfatah.ac.id

M. Amer et al. (eds.), *Territorial Landscapes Solutions for Sustainable Energy and Food Security*, Sustainable Landscape Planning and Natural Resources Management, https://doi.org/10.1007/978-3-031-82419-7_3

system. SDGs have become a shared framework for stakeholders to achieve sustainable shared prosperity. By the five SDGs principles, namely, people, planet, prosperity, peace, and partnership (Mawonde & Togo, 2021), universities have an essential role in achieving the SDGs. Campuses can become centers of excellence in the scientific field by their core competencies, mainstream SDGs in natural processes and actions in education, teaching, and research, and partner with the central and regional governments and other stakeholders in monitoring and evaluating SDGs implementation.

Through the Tridharma of Higher Education and their contributions to the SDGs in the field of research, universities can take the form of research related to the SDGs, innovation and solutions, interdisciplinary and transdisciplinary research, and increasing the capacity of researchers. The field of education includes education for sustainable development, mobilization of the younger generation, and increasing understanding and capacity. Furthermore, the field of community service includes facilitators and catalysts of public involvement, cross-sector dialogue and action, community advocacy, and increasing stakeholder commitment. Environmentally friendly behavior, responsible consumption, and production must be implemented starting in the campus environment; green campus innovation can be an inspiration and action for the entire campus community so that it can be duplicated in daily activities and community life.

Implementation through campus activity programs to create a sustainable green campus is essential to be carried out in an integrated manner with the tri dharma of higher education. The area of campus B (Jakabaring) of Universitas Islam Negeri Raden Fatah Palembang (UINRFP) is in swampland (Lussetyowati, 2017). Adaptation is needed to construct campus facilities so that the groundwater conservation function is carried out with the productive function of campus activities. The implications of swampland, areas in urban centers, and for campus development purposes, it is necessary to manage potential flooding (water overflow) in high rainfall, water containment and conservation (building water retention), adequate water storage capacity, building construction technology campus, rainwater harvesting system, building-open area ratio, facilities and domestic waste management (sewage treatment plant), water pollution control, and potential for developing the usability of water resources. The swampland adaptation-based management mentioned above is in synergy with the 6th SDGs for Clean water and sanitation, the 9th SDGs for Industry, Innovation, and infrastructure, the 11th SDGs Sustainable cities and communities, the 13th SDGs Climate action, and the 17th SDGs Partnerships for the goals (Mawonde & Togo, 2021).

Sustainable development in the swamplands of the Jakabaring area (South Sumatra) based on swampland adaptation makes UINRFP's contribution towards carbon–neutral Indonesia 2060. Adaptation efforts in swampland become a middle ground so that all interests, namely conservation and reclamation, can work together. The swampland adaptation-based strategy on the UINRFP campus is an active implementation of SDGs, real action for reducing GHG emissions, and groundwater conservation. Therefore, this paper will examine the performance of various swampland adaptations through (1) rainwater harvesting, (2) infiltration and water storage structures under campus buildings, (3) controlling water pollution and water reserves from recycling, and (4) management of campus area water resources.

2 Materials and Methods

2.1 Location and Research Methods

The research was conducted in the Jakabaring Campus (Campus B) area of UIN Raden Fatah Palembang in Palembang, South Sumatra Province (−3.010S; 104.773E), with an area of 15 ha. The research method uses primary data sources and secondary data. Primary data was obtained from field observations, field data measurements, photo documentation, and interviews. Secondary data uses Detail Engineering Design (DED) data, namely As-Built-Drawing Mechanical, Electrical, Structure, Infrastructure, and Hardscape and Softscape (Islamic Development Bank-UINRFP-PT Nindya Karya (Persero)-PT Deta Decon, 2018).

The research objects are the campus land area, seven units of main buildings (administrative headquarters, library, integrated laboratory, Faculty of Science and Technology, Faculty of Economics and Islamic Business, Faculty of Tarbiyah and Teacher Training, and Faculty of Psychology), campus utilities (sewage treatment plant, water treatment plant, ground water tanks 1 and 2, powerhouse, and ground tank (sump-pit)).

2.2 Data Types and Analysis

Building data uses DED data. Data on the number of active Jakabaring campus students, lecturers, and education staff from secondary data in the personnel section. Monthly rainfall data for ten years (2013–2022) was obtained from BMKG Class I Kenten Palembang (BMKG, 2023), then the mainstay (adequate) rainfall will be determined. Water quality tests for canals, retention ponds, and STP outlets are carried out at the UINRFP Integrated Laboratory on variables according to the respective water and wastewater quality standards according to Government Regulation 22/2021 and Minister of Environment and Forestry Regulation P.68/2016 (Indonesia, 2021; KLHK, 2016).

Reliable rainfall is needed to calculate the discharge from a reliable water source for a particular purpose. The mainstay discharge for this research was used as a criterion of $\geq$80% chance of being used as the mainstay rainwater. This functions as water availability and is the campus area's water potential, which is used for lecture activities outside of drinking water. The mainstay rainfall probability value is calculated using the formula:

$$P(\%) = m/(n + 1) \times 100\% \quad (1)$$

where $P\%$ is the probability of reliable rainfall, m is the data sequence, and n is the number of data. Analysis of the potential for rainwater harvesting for each building is used to determine the quantity of rainwater based on the amount of rain that falls and is caught by the building's roof. Calculated by the formula:

$$\text{RWH} = R \times La \times Rc/1000 \quad (2)$$

where RWH is the monthly rainwater harvest (m^3), R is the monthly rainfall (mm), La is the roof area (m^2), and Rc is the building roof runoff coefficient 0.9 (Suripin, 2004). Analysis of water needs is calculated based on the average number of academics and educational staff (according to the water requirement standards of the Directorate General of Human Settlements, Ministry of Public Works (2000), namely the standard for clean water requirements for educational facilities of 10 L/person/day (Mulawarman et al., 2021).

3 Results and Discussion

3.1 Rainwater Harvesting System

The rainwater harvesting system is designed to manage water resources on campus because the campus area is swampland with excessive water volume. The Rainwater Harvesting System is intended to reduce the amount of water reaching the ground surface and prevent waterlogging/flooding. Rainwater is collected and channeled into canals within the campus as a backup water source (Zhang et al., 2009). Water reserves need management to achieve water quality and quantity so that it is suitable for meeting the basic water needs of life (Amalia et al., 2023).

Rainwater harvesting is a strategy for managing water resources in swampland, making efforts to recycle water, and providing water reserves for the needs of the campus and surrounding areas. Water reserves from Rainwater Harvesting are stored in water canals within the Jakabaring Campus area, with several goals and water conservation efforts such as (1) providing water throughout the year, (2) watering plants, (3) cleaning campus areas and vehicles, (4) raw water in hydrants for fire control, (5) potential for fisheries production, (6) potential for bottled drinking water (AMDK) production as a business unit, (7) potential for micro-hydro renewable energy, etc.

The buildings on the Jakabaring Campus of UIN Raden Fatah Palembang (UINRFP) include seven central units, four existing supporting buildings, and one under construction. Rainwater harvesting is implemented in 7 main buildings, namely: the central administration building (Rectorate), library, integrated laboratory, faculty of science and technology (FST), faculty of Islamic economics and business (FEBI), faculty of tarbiyah and teacher training (FITK), and faculty psychology (FPsi). Rainfall data for Palembang 2013–2022 is the central (adequate) annual rainfall of 2354 mm (BMKG, 2023; Mulawarman et al., 2021); the rainfall distribution per month is in Fig. 1. Of the seven main campus buildings built for harvesting facilities, The annual volume of rainwater obtained is 30,424 m^3, from the building and rafter area of 14,359 m^2 (9.57% of the Jakabaring campus area). The rainwater harvest value was excellent at FST and FEBI and lowest at Rectorate, respectively 5633 and 2905 m^3 (Fig. 2).

From rainwater harvesting for a year at the Jakabaring campus, 30,424 m^3 can meet the campus average clean water needs of 28,653 m^3. There was a water surplus of 7958 m^3 in 5 months but a water shortage in the other seven months (6187 m^3), especially during the dry season from May to October (Table 1). Based on existing data, the campus area can meet its needs independently if the maximum rainwater harvesting system is managed and operational from all main buildings in the campus area (Fig. 3). These results provide scientific value in efforts for real action for water conservation and optimal management. They are a form of implementation of the 6th SDGs Clean water and sanitation in water resource management in the UINRFP campus area.

Rainwater harvested from the roofs of all buildings is then channeled to the base area of the building and collected into the central water canal surrounding the campus area through a watertight channel. Water canals accommodate and regulate the hydrological system, so water does not inundate the land. Rainwater harvested and stored in water canals is then used for watering plants, cultivating freshwater fish, the primary water source for fountain fish ponds, and controlling the supply and management of water widely in the campus area.

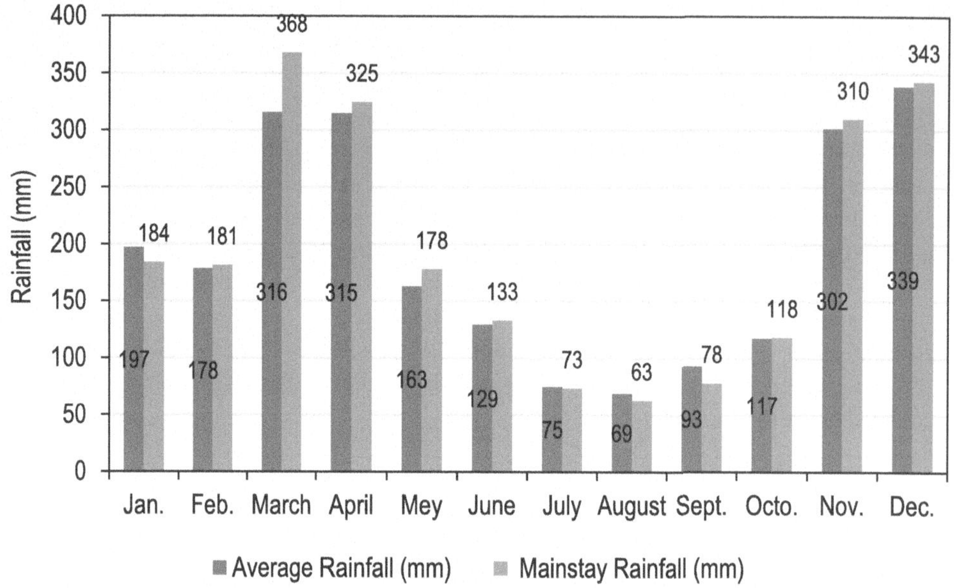

Fig. 1 Average monthly rainfall and mainstay rainfall

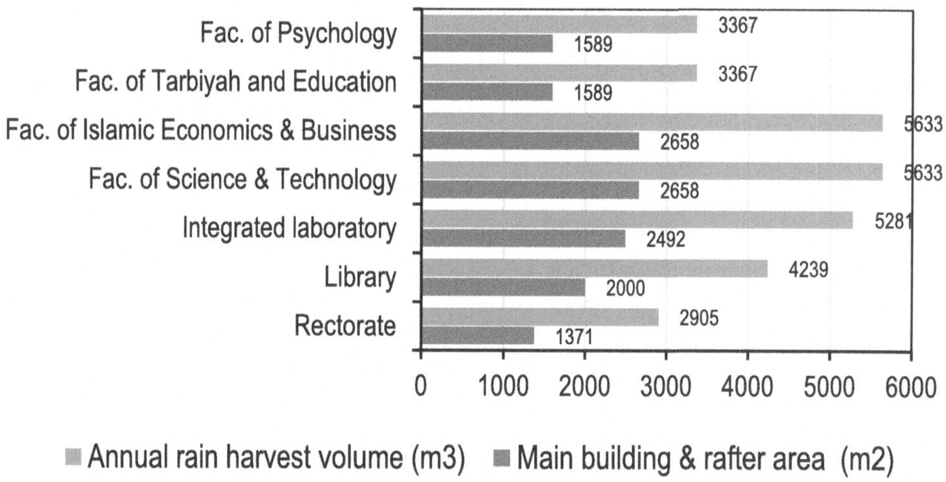

Fig. 2 Roof and rafter area of the building and volume of rainwater harvesting

Table 1 Rainwater harvesting and clean water needs in the campus area

Month	Mainstay rainfall (mm)	Building roof-rafter area (m^2)	Rainwater harvesting (m^3)[a]	Number of days	Water requirement (m^3/month)[b]	Excess water (m^3)	Water shortage (m^3)
Jan	184	14,359	2378	31	2434		56
Feb	181	14,359	2344	28	2198	146	
March	368	14,359	4754	31	2434	2321	
April	325	14,359	4195	30	2355	1840	
May	178	14,359	2299	31	2434		135
June	133	14,359	1719	30	2355		636
July	73	14,359	949	31	2434		1485
August	63	14,359	814	31	2434		1619
Sept	78	14,359	1007	30	2355		1348
Oct	118	14,359	1526	31	2434		907
Nov	310	14,359	4007	30	2355	1652	
Dec	343	14,359	4432	31	2434	1999	
Amount	2354		30,424	365	28,653	7958	6187

[a]According to Suripin (2004), the runoff coefficient for roofed areas is 0.75–0.95 (0.9)

[b]The standard for clean water requirements in educational facilities is 10 L/person/day (Directorate of Human Settlements, Department of Public Works (2000) in Mulawarman et al., 2021); 7850 people (Jakabaring Campus)

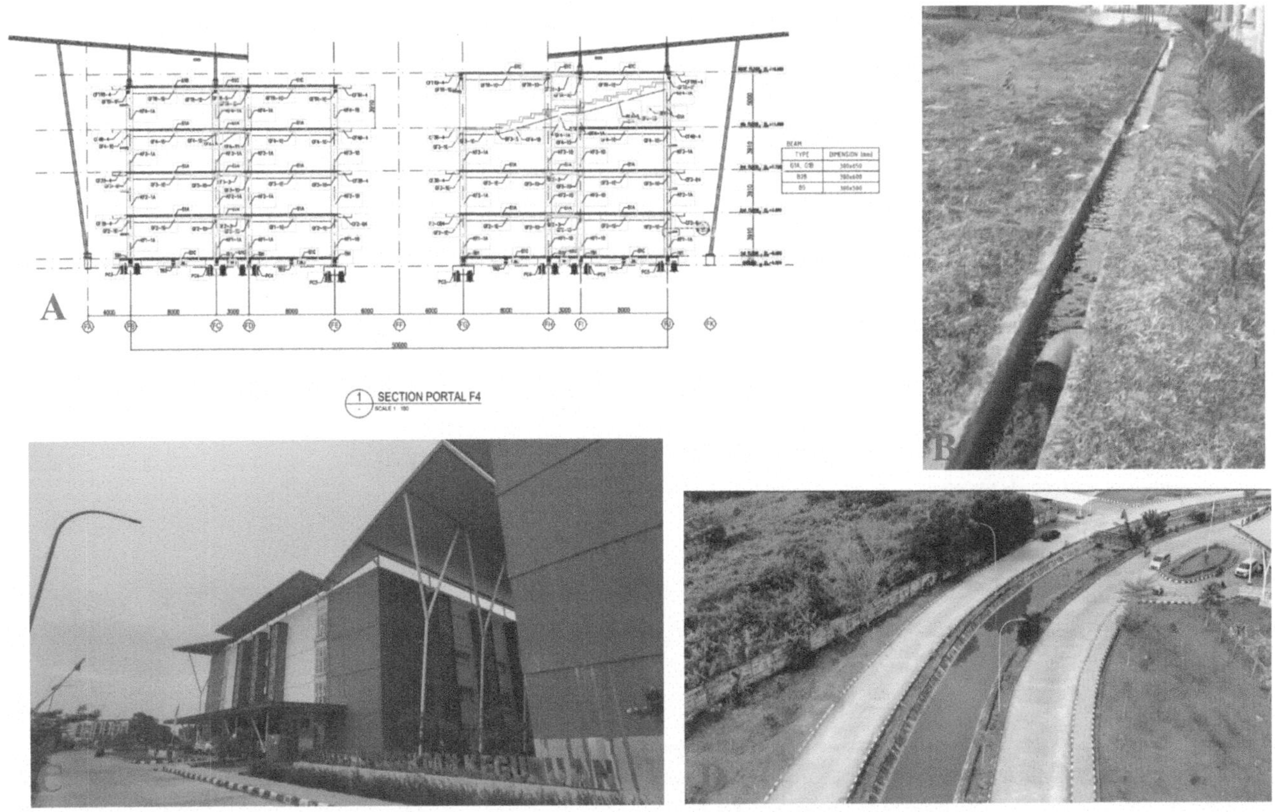

Fig. 3 Rainwater harvesting operations, **a** technical drawing of the building's roof and rafters; **b** the flow of harvested rainwater towards the water canal (closed system); **c** existing roof and rafter shape; **d** water canals to accommodate rainwater harvesting

3.2 Infiltration and Water Storage Under Buildings

The swampy land in the campus area is part of Palembang City, which has a history as a maritime kingdom, with swamps surrounding 50% of the topography of Palembang City. So the Jakabaring UINRFP campus area was designed using a square pile type precast concrete foundation so that the ground floor of the building does not rest directly on the ground surface; this makes the bottom of the building floor have an additional water storage and absorption area (cavity).

The seven main buildings are supported by this 1321 healthy foundation system (1221 are under the building area, and the other 100 are on the terrace and sides of the building). The floors of the building are made of cast reinforced concrete, including the first floor. Water that comes from rainwater fills and is stored in the cavity naturally and becomes an effort to collect and conserve water by maximizing the absorption and storage of rainwater. The volume of the cavity under the building is estimated to reach 5165.4 m^3, and the assumption is that 65% will be filled with 3357.5 m^3 of water (from a building area of 9519 m^2 or 6.35% of the Jakabaring campus area) (Table 2). This amount reaches > 11% of the total rainwater harvest in the campus area. This stored water will be infiltrated below the soil's surface by gravity, according to local environmental conditions, including water saturation level, soil type and density, rainfall, drainage, and leaks in the surrounding land system (Djamaluddin et al., 2020).

The water cavity under the building increases the amount of water stored in the campus area throughout the year. Apart from that, it is a breeding habitat for swamp wildlife that previously existed in the area, including types of swamp fish such as snakehead (Channa striata), betook (Anabas testudineus), and selling (Silver rasbora) (Adamson, 2010). Water resources that are conserved in the basement of this building can help prevent flooding of campus land and, at the same time, become a water reserve for life ecologically and economically. The ecological-hydrological function of swamp land can occur well through this technique in the 6.35% (area under campus buildings), namely the availability of water on the land throughout the year (cavities under buildings) (Fig. 4) for both. Water and soil conservation functions and conservation of swamp water flora and fauna.

Table 2 The volume of water storage under the main campus building

Building	Building area (m^2)	Number square pile in the building	Net under building area (m^2)	Average cavity height under the building (m)[a]	Water storage volume (m^3)
Rectorate	1128	153	1103	0.45	496.5
Library	1805	214	1771	0.65	1151.2
Integrated laboratory	1530	218	1495	0.65	971.8
Fac. of science and technology	1616	212	1582	0.55	870.1
Fac. of Islamic economics and business	1616	212	1582	0.55	870.1
Fac. of tarbiyah and education	912	106	895	0.45	402.8
Fac. of psychology	912	106	895	0.45	402.8
Amount	9519	1221	9324	–	5165.4
Water conservation (65%)					3357.5

[a]Average length—the distance between the concrete base of the 1st floor and the ground surface

Fig. 4 Space under the building as water conservation (reservoir and water absorption media). **a** Library building, **b** integrated laboratory

3.3 Control of Water Pollution and Water Reserves from Recycling

All wastewater produced from the campus area is collected in ground-tanks located around the building and further processed using a sewage treatment plant (STP) or integrated wastewater treatment plant (IPALDT) (Fig. 5). From seven buildings, wastewater is collected in the nearest ground tank and sent via relay to STP for further processing. There are 25 ground tanks with a capacity of 2–21 m^3 and a total capacity of 166 m^3 (operational volume 149 m^3) (Table 3). With an average attendance of academics at the Jakabaring campus of 5495 people per day (70% of 7850 people), around 46.7 m^3/day of domestic wastewater is produced (Table 1). Domestic wastewater production is processed at an STP with an installed capacity of 506 m^3, with a hydraulic retention time (HRT) of 10.8 days.

Wastewater treatment operations at STP are carried out using a combined physical–chemical–biological method, including filtration, coagulation, flocculation, aeration, and activated sludge. The five methods are applied simultaneously and adjusted to the quality of the wastewater inlet to achieve an STP outlet that meets domestic wastewater quality standards based on the Minister of Environment and Forestry Regulation P.68/2016 (KLHK, 2016). STP outlet quality monitoring is carried out periodically to meet water quality standards (except pH in retention pond 1) and wastewater so that it can be released into the environment and for further use (Table 4).

Efforts to recycle domestic wastewater were developed to utilize treated wastewater by channeling it into retention ponds for reserve water. The retention pond also becomes a bio-control and bio-indicator pond with the cultivation of local fish species. The retention pond that holds reserve water has an area of 4550 m^2 and a water volume of 5314 m^3. Thus, the treated wastewater the retention pond receives daily is 0.88% of its capacity. The volume of this retention pond and the existing annual rainfall can store and assimilate STP outlet wastewater.

3.4 Potential Water Resources in the Campus Area

Water resources are an essential part of the source of life, including the tri-dharma activities of higher education on campus. Water is a renewable resource through a cyclical mechanism on earth. Other resources are needed to

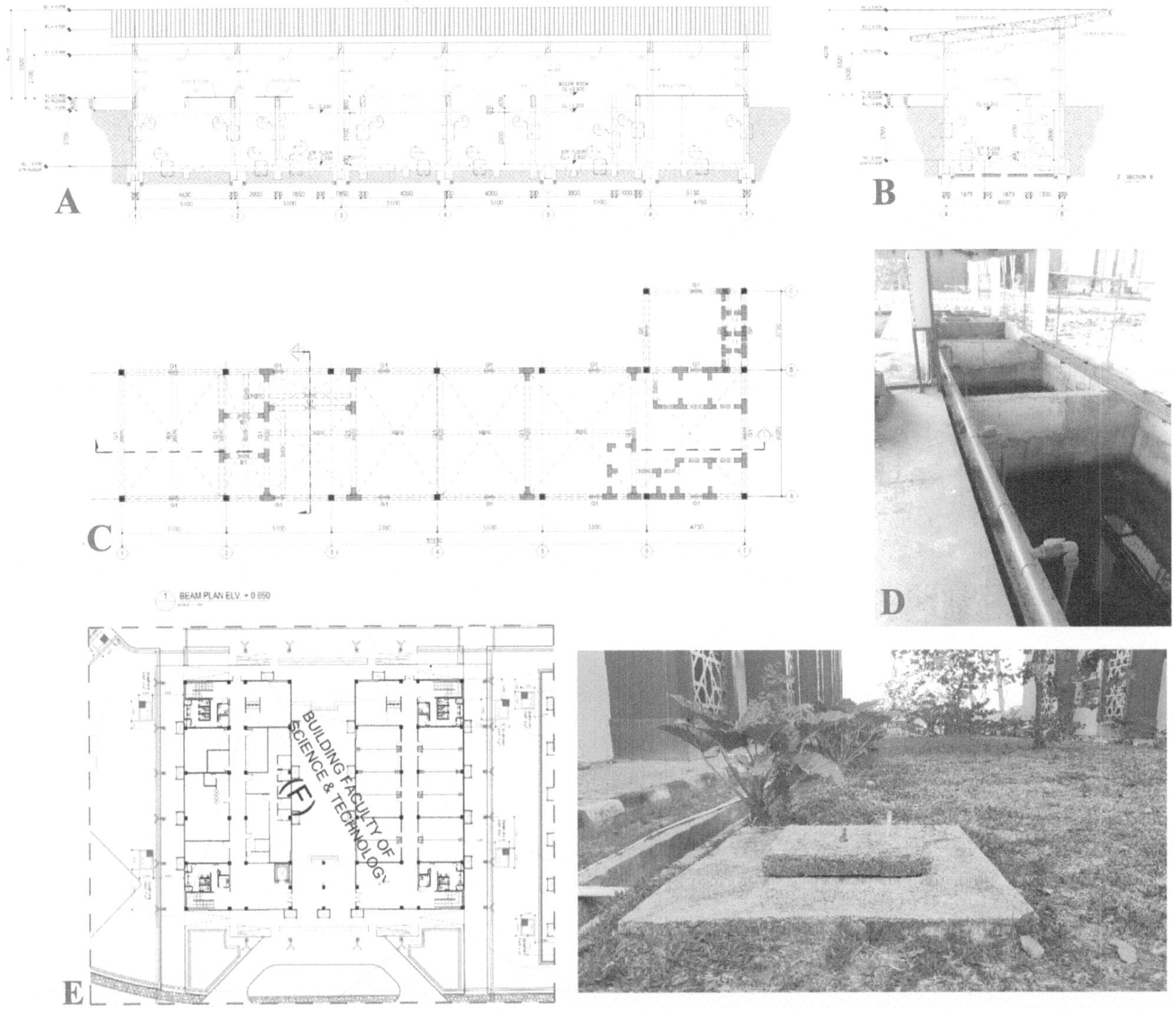

Fig. 5 Sewage treatment plant (STP) and ground tank, **a**, **b**, **c** technical drawing of STP, **d** existing STP capacity 506 m^3, **e** technical drawing and location of ground tank, **f** existing ground tanks

meet the quantity and quality of water. The worse and even more polluted it is, the greater the costs and higher the greenhouse gas (GHG) emissions for its management (Muhammetoglu et al., 2023).

Swampland on the UINRFP campus requires optimal water resource management so that water resources become efficient and effective. Rainwater harvesting is done to prevent waterlogging. The roof area of the building was even widened (with rafters) (increased by 50.84%; from 9519 to 14,359 m^2) to increase the volume of rainwater catchment obtained and minimize rainwater reaching the ground surface, thereby minimizing the formation of water runoff and no puddles occurring in the campus area.

From rainwater harvesting, rainwater catchment from water bodies, and springs in water canals, the first retention pond (in front of the Faculty of Psychology), the second retention pond (in front of the Administrative Headquarters/Rectorate) obtained a discharge or volume of 75,171 m^3/year (Table 5) (Fig. 6). This potential water resource needs management and investment so that it has sustainable economic, ecological, and social benefits. The swampland adaptation approach can take place optimally with the primary objective: control of swampland as a productive operation for the campus and groundwater conservation. Further future development will carry out water-based businesses sufficient for internal needs or external business units.

Various productive efforts based on swampland adaptation are efforts to preserve swampland so that it remains functional in a sustainable manner for life, especially the function of swampland for conservation and water

Table 3 Number and capacity of ground tanks for domestic wastewater treatment

Building	Number of ground tanks (unit)	Ground tank capacity (m^3)	Total capacity (m^3)
Rectorat	2	5	10
Library	2	5	10
Integrated laboratory	5	2	10
	1	7	7
	1	14	14
Fac. of science and technology	4	7	28
Fac. of Islamic economics and business	4	7	28
	1	14	14
Fac. of tarbiyah and education	2	6	12
Fac. of psychology	2	6	12
	1	21	21
Amount	25		166
Amount (ratio to wastewater 0.9)			149

management, conservation of flora and fauna, food and nutrition sources for living creatures, prevention of pollution and pollution. Environment and fisheries production (nutrient-protein) (Khodijah et al., 2023). The existence of water bodies (water canals and retention ponds) within the campus area is an action to adapt swampland, which has many benefits apart from its function as water management, home to aquatic biota and fisheries, and also has potential as a renewable an energy resource (Khodijah et al., 2023). Renewable energy based on hydropower has the potential for the development of micro-hydro power plants (MHPs). Utilizing water canals as reservoirs for rainwater harvests for water reserves, the existence of retention ponds and recycling efforts in the use of water in swamp areas, as well as efforts to develop MHPs, will be the implementation of the 13th SDGs goal, namely action on climate in the aspect of water resource management.

4 Conclusion

The campus's contribution towards achieving Indonesia's 2060 carbon–neutral target is real action to implement SDGs. Management of swamp-based water resources in the campus area is the implementation of the 6th SDG, Clean Water and Sanitation, and the 13th SDG, Climate Action. The building construction is designed to harvest rainwater. Rainwater harvested in the campus area of 30,424 m^3/year is channeled and stored in water canals surrounding

Table 4 Laboratory test results for the quality of pond water, canal water, and STP wastewater

Variable	Unit	Retention pool water	Canal water	Outlet STP	Standart Gov. Reg 22/2021	Standart Reg. of MEF P.68/2016
Physics						
Smell	–	No smell	No smell	–	–	–
Total dissolved solid (TDS)	mg/L	422	109	124	1000–2000	2000–4000
Chemistry						
pH	–	3.21[a]	7.82	7.82	6–9	6–9
Oil/Fat	mg/L	4.60	1.76	–	5	–
Dissolve oxygen (DO)	mg/L	2.16	1.84	1.92	1–6	–
Electric conductivity (EC)	jus/cm	840	208	246	–	–
Oxidation reduction potensial (ORP)	mV	230	−53.1	−58.4	–	–
Salinity	‰	0	0	0	–	–
Pb	mg/L	<0.004	<0.004	<0.004	0.03	–
Cd	mg/L	0.005	0.005	<0.004	0.01	–
Cu	mg/L	<0.004	<0.004	<0.004	0.02	–
Fe	mg/L	0.248	0.304	<0.004	0.3	–
Biology						
Fecal coliform	MPN/100 mL	<3	1100	1100	100–2000	–
Total coliform	MPN/100 mL	93	1100	1100	1000–10,000	3000

[a]The test results do not meet quality standards

Table 5 The volume of potential water resources in the campus area

Water body	Area (m^2)	Length or height (m)	Water volume (m^3)	Water discharge (m^3/yr)
Water canal[a]	6.05	940	5683	53,993
Retention pond 1[b]	4550	1.6	5314	18,596
Retention pond 2[b]	427	1.6	499	2582
Amount	–	–	11,495	75,171

[a]The water channel area is a cross-section; water reserves come from rainwater harvesting, rainwater capture, and springs
[b]Water sources come from rainwater catchments and springs

the campus area. Construction techniques for campus buildings and utilities with precast concrete pile foundations (The Cyclops concrete type square pile) with a depth of 32–40 m reaching 1461 units (in Building 1221 units). This construction technique is an adaptation solution for urban swampland, estimated to be able to conserve 3357.5 m^3 of water. Domestic wastewater is collected in 25 ground tanks (capacity 2–21 m^3) and channeled to STP for further processing. After meeting the quality standards, it is channeled into the retention pond to become campus operational reserve water. Water resources and their management at UINRFP are water reserves of 75,171 m^3/year with the potential for micro hydro power generation, clean raw water, fisheries production, campus drinking water production, and development into productive business units.

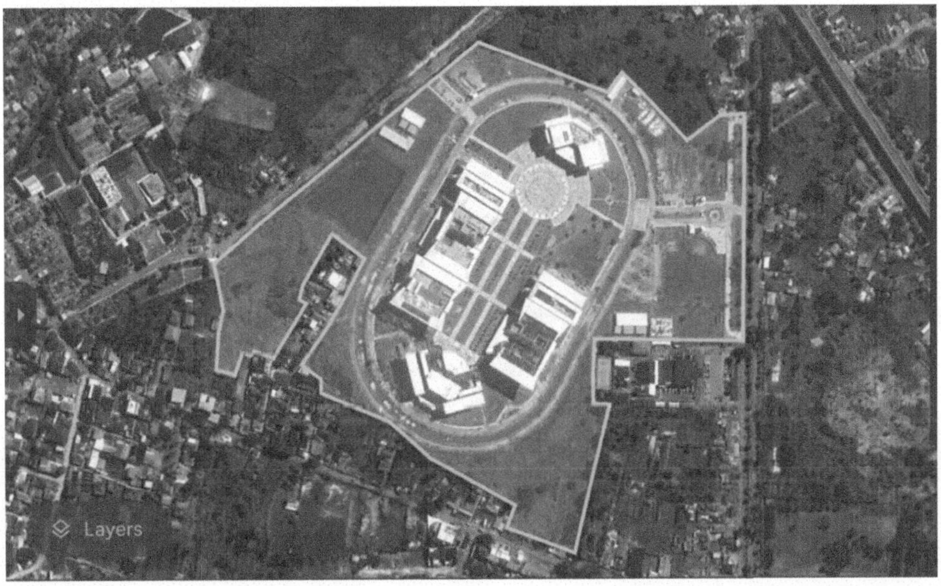

Fig. 6 Conservation of water resources with surrounding water canals and retention ponds 1 and 2 on the Jakabaring Campus area

References

Adamson, E. A. S. (2010). *Influence of historical landscapes, drainage evolution and ecological traits on patterns of genetic diversity in South East Asian freshwater snakehead fishes*. PhD thesis submitted to the Queensland University of Technology, Brisbane, Australia, 241 pp.

Amalia, V., Sanjoto, T. B., & Hardati, P. (2023). Distribution of springs and household clean water needs in Bulu district, Rembang Regency. *International Journal of Research and Review, 10*(9), 44–53.

Badan Meteorologi dan Geofisika (BMKG). (2023). *Data Iklim dan Curah Hujan Palembang*. Stasiun Klimatologi Kelas II Kenten Palembang. Palembang.

Djamaluddin, I., Aly, S. H., Rahim, I. R., Zubair, A., Ibrahim, R., & Abdullah, N. O. (2020). Pengelolaan Drainase Kota Sebagai Upaya Mitigasi Banjir Kota Makassar. *Jurnal Tepat (Teknologi Terapan Untuk Pengabdian Masyarakat), 3*(2), 98–112.

Indonesia. (2021). Peraturan Pemerintah Republik Indonesia Nomor 22 Tahun 2021 tentang Penyelengaraan Perlindungan dan Pengelolaan Lingkungan Hidup. Republik Indonesia: Jakarta.

Kementerian Lingkungan Hidup dan Kehutanan (KLHK). (2016). Peraturan Menteri Lingkungan Hidup dan Kehutanan Republik Indonesia Nomor: P.68/Menlhk-Setjen/2016 tentang Baku Mutu Air Limbah Domestik. Kementerian Lingkungan Hidup dan Kehutanan: Jakarta.

Khodijah, N., Putro, L. H. S., Hadi, A., Aljabar, J. L., & Ichsan, C. (2023). Solar electricity energy: Utilization of renewable energy sources to realize a sustainable campus at UIN Raden Fatah Palembang. *Journal of Sustainability Perspectives, 3*(2), 209–217.

Lussetyowati, T. (2017). Local wisdom in the formation of cultural landscapes in wetland settlements in Palembang City. *Jurnal Arsitektur dan Perkotaan "KORIDOR", 8*(2), 138–143.

Mawonde, A., & Togo, M. (2021). Challenges of involving students incampus SDGs-related practices in an Odel context: The case of The University of South Africa (Unisa). *International Journal of Sustainability in Higher Education, 2021*(1), 1–16.

Muhammetoglu, A., Al-Omari, A., Al-Houri, Z., Topkaya, B., Tumbul, T., & Muhammetoglu, H. (2023). Assessment of energy performance and GHG emissions for the urban water cycle toward sustainability. *Journal of Water and Climate Change, 14*(1), 223–238.

Mulawarman, A., Munaja, R., Limung, K. (2021). Potential for rainwater harvesting to fulfill clean water needs in the University of West Sulawesi Education Area. *Bandar: Journal of civil Engenering, 3*(2), 29–36.

Suripin. (2004). *Sistem Drainase Perkotaan yang Berkelanjutan*. Yogyakarta, Indonesia: ANDI Offset.

Zhang, Y., Chen, D., Chen, L., & Ashbolt, S. (2009). Potential for rainwater use in high-rise buildings in Australian cities. *Journal of Environmental Management, 91*, 222–226.

Examining Innovation Agility: A Review of the Bio-based Business Models

Siti Yasmina Zubaedah, Avanti Fontana, Alexander Tifaona and Posma S. J. Kennedy

Abstract

Drawing from biology, sustainability means "survival of the fittest," businesses that can evolve in line with the ecosystem's natural trajectory will live on while those who fail will cease to exist. Business survival requires building the agility to effectively evolve, continuously capture value, and grow. This study is to examine the agility of current bio-based business models using a framework derived from the basic premise of effective business model evolution in order to identify bio-based business model innovation opportunities. To ensure proper examination, this article focuses on two bio-based businesses, which are Styrax Benzoin and Pongamia Pinnata. Using the prescribed business model agility lens, the method of analysis consists of examining the Styrax business model and the Pongamia business model. Analysis found that current practices of producing styrax benzoin resins remain unchanged, which indicates the pressing need for developing an innovative business model and building agility. Although still in the development stage, the Pongamia case offers an alternative solution with the cultivation of Pongamia seeds as the backbone for the bio-energy business model. Overall, embedding sustainability principles in formulating the bio-based business model is pertinent for ensuring agility. The agility analyses resulted in the identified prerequisites for the formulation and execution of bio-based business model innovations, namely, human capital, governance, and sustainability principles. Future studies using the agility framework allows for the development of business model standards aimed at improving quality, increasing quantity, and overall maximizing value of bio-based products.

Keywords

Business model agility · Innovation · Natural resources · Sustainability

1 Introduction

Sustainability necessitates the synchronization of strategic considerations in the logical (rationalism), ecological (naturalism), and sociological (humanism) aspects (Senge & Carstedt, 2001). Any issue that may impact the business' circumstances in terms of its economics, ecology, and stakeholders must be considered when formulating a new strategy. Regardless of the environmental changes, management plays the determining role of selecting the choices that synchronizes firm particulars with the surrounding circumstances. In order to sustain performance, firms need to adapt effectively by consistently adjusting strategy, structure, and process to the environment (Miles & Snow, 2003). Adaptability requires the ability to create knowledge based on the information captured from the environment in order to ensure strategy formation leads to performance (Zubaedah, 2022), as misreading the environment would lead to the selection of the wrong strategy or failure to build capabilities required for firms to survive (Reeves et al., 2016). Drawing from biology, sustainability means "survival of the fittest," businesses that can evolve in line with

S. Y. Zubaedah (✉)
Universitas Bunda Mulia, North Jakarta, Indonesia
e-mail: szubaedah@bundamulia.ac.id

A. Fontana
Faculty of Economics and Business, Universitas Indonesia, Depok, Indonesia
e-mail: avanti.fontana@ui.ac.id

A. Tifaona
PT Lembata Hira Sejahtera, East Nusa Tenggara, Indonesia
e-mail: lembata.sejahtera@gmail.com

P. S. J. Kennedy
Faculty of Economics and Business, Universitas Kristen Indonesia, East Jakarta, Indonesia
e-mail: posmahutasoit@uki.ac.id

M. Amer et al. (eds.), *Territorial Landscapes Solutions for Sustainable Energy and Food Security*,
Sustainable Landscape Planning and Natural Resources Management, https://doi.org/10.1007/978-3-031-82419-7_4

the ecosystem's natural trajectory will live on while those who fail will cease to exist.

The objective of this study is to examine the agility of current bio-based business models using a framework derived from the basic premise of effective business model evolution to identify bio-based business model innovation opportunities. As businesses go through adaptation cycles, agility can be observed from how well the organization is able to manage the changes in the business model over some period of time, which is denoted as Business Model Effectiveness (Zubaedah & Fontana, 2023). To ensure proper examination, this article focuses on two bio-based businesses, i.e., Styrax Benzoin and Pongamia Pinnata.

2 Literature Review

The following literature review encompasses the theoretical foundations of the study framework, which is an extension of previous works related to strategic agility (Zubaedah, 2016; Zubaedah & Fontana, 2023).

2.1 Business Model and Sustainability

As a construct, the business model is a powerful analytical tool shown to be most useful for evaluating firm performance (Shafer et al., 2005). Examining sustainability principles in a business model provides many insights as the organization's value creation logic as well as how the business impacts the surrounding community are clearly articulated in the business model design, so as providing opportunities to develop new governance forms for new partnerships or cooperations (Schaltegger et al., 2016). Participation of private sectors in developing natural environment and society is central in corporate sustainability (Schaltegger & Burritt, 2005). A business model design can be formulated to take into account ecological, social and economic values simultaneously aimed at sustainability. Stubbs and Cocklin (2008) concluded that a sustainability business model embeds internal structural and cultural capabilities as well as build collaborations with related stakeholders so as to not only attain sustainability at the level of the firm but also the society.

2.2 Business Model Innovation

A business model innovation can be characterized as adopting new activities, implementing new ways to connect activities, or altering the parties conducting the activities (Amit & Zott, 2012). Throughout adaptation cycles, a firm will arrive to the point where existing operations can no longer prevail and adjustments in the business model must be undertaken (Zubaedah, 2016). In contrast with other types of innovations, a business model innovation requires explorations and needs to be conducted with full engagement from top management (Hossain, 2017). Consequently, effectively managing one business model innovation after another as the firm goes through adaptation cycles require effectively managing those changes (Zubaedah & Fontana, 2023). Despite the barriers to organizing changes, business model innovation is vital for maintaining good performance and any difficulties can be overcome with experimentations (Chesbrough, 2010).

2.3 Agility and Sustainability

Building strategic agility requires prioritizing speed over perfection, flexibility over planning, diversification and "efficient slack" over-optimization, empowerment over hierarchy, learning over blaming, and resource modularity over lock-in (Wade et al., 2021). There is a growing number of studies emphasizing the significance of including sustainability elements for innovating the business model (Hossain, 2017). Elements related to emphasized considerations on the interests of all stakeholders, including the environment and the communities, would lead corporate initiatives towards creating competitive advantage and sustainability (Baumeister et al., 2015). Incorporating sustainability factors in the business model innovation has shown to be the source of strength for non-profit organizations as well as for elevating the positioning of global corporations (Aagaard & Lindgren, 2015). As companies evolve, changes in the business ecosystem will require adjustments in the business model where sustainability considerations are pertinent to ensure effective adaptation.

3 Theoretical Framework

The three-lens agility framework (Zubaedah & Fontana, 2023) consists of the Business Model Effectiveness lens, Productive Information System lens and the Independent-Dynamic Culture lens. Considering that most bio-commodities production in Indonesia are managed using traditional (non-corporate), agroforestry model, examining agility can sufficiently be attained using the business model lens as presented in Fig. 1.

3.1 The Business Model Agility Lens

Previous studies prescribed the business model framework to comprise multiple components with three common

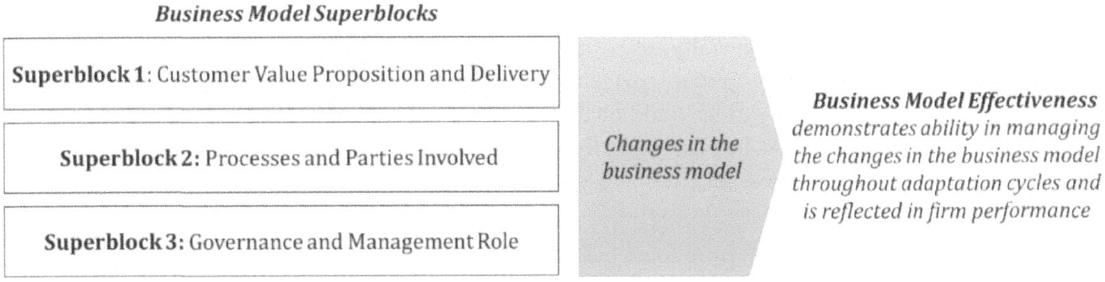

Fig. 1 Agility framework—business model effectiveness lens

themes, which will be denoted in this article as business model superblocks. The first superblock is Customer Value Proposition (CVP) and delivery that describes the value offered and how value is delivered to the identified intended customers. The second superblock describes the processes and parties involved in value offered from products (goods or services). The third superblock includes elements related with governance and management role. These three superblocks merge together to form the company's business model.

Based on the basic premise that the firm's responses towards environmental changes are reflected in the adopted business model (Fontana & Zubaedah, 2012), agility can be observed from how well the firm's business model evolves given the demands of the ecosystem. As firms go through natural cycles of adaptation, agility is presented in the ability of managing business model innovations that result in consistently good performance, which is denoted as Business Model Effectiveness (Zubaedah & Fontana, 2023). In other words, Business Model Effectiveness becomes the proxy for examining agility.

3.2 Method of Analysis

Case analyses were conducted through narratives describing the business models examined based on data and information collected from scientific literature, online publications, and interviews with practitioners or business actors. In the following discussions, the bio-based products are explained to provide context of the cases analyzed in this article. Using the prescribed business model agility lens, the method of analysis adopted for this study consists of examining the Styrax business model and Pongamia business model.

The Styrax business model was analyzed based on the combined information obtained from literature review and direct observations of the traditional cultivation practices. In addition to literature review, the Pongamia case was analyzed based on the business practices of PT Lembata Hira Sejahtera, which is a newly established company pursuing bio-energy production from Pongamia oil as feedstock. The two cases under study share a common thread of businesses that are derived from bio-commodities. Once a baseline mapping of the business model evolution of these cases was completed, the analyses provide deeper insights through compare and contrast between the two models.

4 Case Analyses

4.1 Bio-based Products

Indonesia is ranked number two as the most biodiverse country in the world (World Atlas, 2023). Considering the extensively diverse bio-based commodities, Indonesia has the opportunity to strengthen its positioning in the growing industries related with bio-based products.

4.2 Styrax Benzoin

In this article, the term Styrax refers to the tree while Styrax benzoin denotes the fragrance resins produced from the tree saps secreted from wounded tree barks. Styrax is a genus of over 100 species of trees that can produce a fragrance resin called benzoin, which is categorized under the phenolic resin containing benzoic acid, cinnamic acid, and their derivatives (He et al., 2023). Since its cultivation efforts in the early 1900s, they are mostly found in the natural forests throughout North Sumatra province, which provides for the production of three main types of resins, namely, Toba (Styrax paralleneurum Perk), Durame (Styrax benzoine dryland), and Bulu (Styrax benzoin var. Hilliferum) (Kiswandono et al., 2016).

Styrax naturally grows in natural forests and is usually under the management of the traditional families of certain clans native to the area. Although the legal status of the land was designated as production or state-owned forests, the communities acquire Styrax from family legacies, handed down from generation to generation (Harada et al., 2022).

In addition to the economic significance, Styraz benzoin is an integral part of the local culture. Generally, uses of Styrax benzoine are characterized into two groups, which are the traditional use as incense used in various traditional ceremonies, and the modern use as a bio-chemical ingredient in pharmaceutical products, perfumes, and cosmetics (Kebun Raya Cibodas, 2021).

4.3 Pongamia Pinnata

Found throughout the Asia, Pongamia pinnata (or "Pongamia," also called Indian Beech or Malapari in Bahasa Indonesia) is known for its many medicinal properties. Widely used in Ayurvedic treatments, use of Pongamia bark as medications for treating wounds and ulcers can be traced back to traditional Sri Lankan medical practices (Balasooriya et al., 2021). Pongamia also have other medical uses including as biopesticide, insect repellent, ingredient in soap production, and as a source of edible grade vegetable oil (Degani et al., 2022).

Flowers from Pongamia tree are rich with pollen and nectar, while Pongamia seeds can be processed into cake then used as bio-energy, food and feed protein, and organic fertilizer (Degani et al., 2022). Various studies have found that Pongamia makes an excellent source for diesel based on Hydrotreated Vegetable Oil (HVO) processing as the seed contains 40% oil (Degani et al., 2022). Through a hydrotreatment process, HVO produced from bio-based feedstock is reported to have lower emissions, higher efficiency for fuel consumption and identical chemical structure with petroleum diesel (Oils and Fats International, 2018).

Newly established in 2023, PT Lembata Hira Sejahtera (or "Batara") discovered the promising opportunity for entering the HVO industry with Pongamia as the main ingredient. Founders of Batara are natives of Lembata Regency, located in the Nusa Tenggara Timur province between East Flores Regency and Alor Regency, and had the knowledge of Pongamia to be one of the indigenous perennial crops. For the local population, Lembata is the natural habitat for Pongamia, which has been used for its medicinal properties.

4.4 Case of Styrax Benzoin

Statistics shows that land area for Styrax in North Sumatra is 23,172 ha in 2021, a slight increase from 23,146 ha in 2020 (BPS, 2023). As reported by the Central Bureau of Statistics (BPS, 2023), Styrax benzoin production amounted to 8845 tons in 2021 and 8604 tons in 2020.

4.5 Agroforestry Business Model

In addition to its cultural significance, Styrax benzoin has been identified as one of the most important Non-timber Forest Products (NTFP) supporting the livelihood of the local communities in the cultivation areas of North Sumatra (Harada et al., 2022). In addition to Styrax, farmers would plant vegetables and fruits as well as such higher-valued plants as coffee and cinnamon (Fernandez, 2004). Traditionally, forest areas with Styrax are referred as "kebun kemenyan," or benzoin gardens, which are plots ranging from 1 to 3 ha and can only be harvested when the trees reach 7–10 years old (Fernandez, 2004).

In 1996–2000, a group of researchers conducted a study to evaluate the overall benzoin system at two villages, which are Kecupak II in Dairi district and Pusuk I in Tapanuli Utara district, in the North Sumatra province of Indonesia (Fernandez, 2004). The two villages were selected in order to compare how benzoin management works between Pusuk who relied heavily from Styrax benzoin for as the economic source, and Kecupak who had other sources of economy. Fernandez (2004) reported that seedlings are collected from the best parental trees available in the existing benzoin gardens. Styrax will reach its productive period between 10 and 60 years old, where tapping, or wounding, the trees is done annually and harvesting is done three times a year after leaving the wounded barks for at least 3 months. Farmers have their own specific criteria of which tree to harvest that year depending on their personal experiences and habits. As a result, the quantity and quality of Styrax benzoin vary between harvests where the overexploited trees easily become weak and susceptible to diseases, which pose a health threat to the entire garden. However, benzoin gardening is no longer considered to be prestigious where the younger generation perceived it to be obsolete. Many have abandoned the Styrax benzoin and cultivated other commodities with promising profits.

Harada et al. (2022) conducted village-level multiple case studies at two locations in the Humbang Hasundutan district, namely, Sosor Tambok with 236.5 ha of forests dominated by Styrax, and Pandumaan with 5100 ha of forest also mainly consist of Styrax. In these two villages, Styrax benzoin has been the main source of local economy for generations. Seedlings were obtained from existing gardens due to the lack of knowledge on creating nurseries. Harvesting resins commence when trees are more than 15 years old, and harvest period is only twice a year. The harvesting process is labor intensive and require many workdays as the barks are firstly wounded and then left alone for three to four months before the hardened resins can be collected. In the two villages under study, there are Styrax owners who inherited the bio-assets from their ancestors and non-owners who work as labor to earn wages. Analysis showed that there is no difference in income from agricultural products between Styrax owners and non-owners. The study concluded that despite the introduction of coffee as an alternative commodity, Styrax benzoin remains the main source of income for the two villages. More importantly, managing and protecting the forests are preserved as a cultural inheritance. To this day, the villages struggle to conserve their way of living from Styrax benzoin despite the price fluctuations that lead to unstable income.

Based on these two previous studies, the Styrax benzoin business model has been preserved to be in accordance with the practices that are passed down from the ancestors to one generation after another. Such circumstances led to the strong sociocultural ties that characterize both Styrax forest management and the Styrax benzoin, which resulted in the survival of this business despite the current stagnant growth and under-optimized performance. Moreover, persistence in referring to past practices results in lack of new knowledge creation and limited solutions. This may pose a threat to the sustainability of Styrax benzoin production and, more importantly, to the product quality.

4.6 Processing and Trading Model

The agroforestry model characterizes the upstream activities of Styrax benzoin production. Post-harvesting processing and logistics chain from the beginning of the downstream activities. Harvested Styrax benzoin is directly sold to the village middlemen who then sell to district traders or dealers with a capacity of 500–2000 kg per month (Fernandez, 2004). Similarly, in the Hambang Hasundutan district, unsorted and wet benzoin harvested are sold directly to the village middlemen who conduct the sorting and drying before re-selling to the traders at the sub-district, which are then sold to the district traders and finally to the local as well as international buyers (Harada et al., 2022). Producers tend to mix the types and qualities of Styrax benzoin, which resulted in an overall lower quality of product and, in turn, lowering the sales price (Kiswandono et al., 2016). Consequently, current production practices resulted in diminishing economic returns for the producers and traders. However, considering the increasing demands for bio-based fast-moving consumer goods, the demands for Styrax benzoin will continue to rise. Nevertheless, developing the existing traditional model solely relies on the quality and quantity harvested at the upstream. This requires further exploration of ways to improve performance on the entire chain, from upstream to downstream.

4.7 Case of Pongamia

Through, in-depth interviews, Founder and Chief Executive Officer of Batara explained that this company was motivated by Pongamia properties for carbon sequestration and as a source of HVO feedstock. Therefore, Batara designed an integrated HVO value chain business model that begins with seed cultivation, followed by a large-scale planting, and ends with HVO production.

4.8 Current Business Model

The current Batara business model is aimed at producing high-quality seeds and seedlings in substantially large quantities. Since Pongamia has not been determined to be a strategic crop for bio-energy feedstock, there are no large-scale plantations in Indonesia. Existing trees grow naturally, available in selected areas of the country, at a limited number, and with various genetic qualities. Batara positions itself to be the pioneer for cultivating Pongamia seeds with various genetic compositions that can perform well in various planting settings. In its business plan, Batara targets production capability that reaches 1 Million seedlings by 2024, then progressively increases production to 50 Million seedlings by 2027.

There are three core activities currently managed by Batara. First, Batara is collecting all genetic variants of Pongamia seeds where Batara is compiling seeds from different islands, provinces, and cities across Indonesia. Second, Batara is developing a 20 ha seed center, which includes the nursery and field-testing area. Third, Batara is preparing for seed export–import for the best Pongamia seed variants. In line with these three main activities, Batara is organizing various community engagement activities to address the socioeconomic needs, while, at the same time, building good rapport with the surrounding communities. A program entitled "Program MAMA PAPA, Menanam Malapari Panen Porang," which is translated as the "Plant Pongamia, Harvest Yam Program" (or denoted as the Program), was just launched in 1 October 2023. In this program, local farmers were requested to plant, upkeep, and maintain Pongamia trees, while the company facilitates the off taking of elephant foot yams planted in their land. Through this program, Batara intends to obtain seeds genetically exclusive from Lembata, or an Elite Lembata Pongamia variant, as part of the research work for producing the best seeds and seedlings.

As Batara founders are descendants of Lembata natives, the family is well-known and highly respected in Lembata.

This circumstance is the main resource for the Company due to ease of coordinating and acquiring support for its operations. With the abundant, undeveloped land area in Lembata, Batara can acquire sufficient area of land to attain its objectives by building cooperation with individuals and government institutions. Moreover, Batara was established based on an alliance with Investancia Group B.V., a Dutch Company who manages more than 200 hectares of Pongamia in Paraguay. The founders have established collaborations with experts in plantation and bio-energy processing as well as experienced strategists and professional consultants. These alliances are another unique resource of Batara where expertise and updated knowledge are easily accessible. In organizing its activities, Batara is supported by an ensemble of subject matter experts and credible practitioners in the bio-energy industry.

In terms of governance, Batara is an Indonesian company with a simple structured organization. Batara's head office is located in Jakarta, while the seed center is located in Lembata. Nevertheless, the simple structure is enhanced with the extended groups of researchers, specialists and consultants. However, current legal status as a small-scale entity limits the capacity of the seeds and seedlings production. Further considerations are necessary to improve the governance and eliminate restrictions.

4.9 Business Model Innovation

Since its initiation, Batara's establishment was intended to be the first implementation stage of the business model innovation undertaken for pioneering Pongamia oil as feedstock for producing good quality HVO at large quantities and Pongamia as the designated plant for land rehabilitation or afforestation. Existing Pongamia trees grow naturally in forests, agriculture fields, and on the coastal areas of Indonesia. Nonetheless, high-quality seeds and seedlings are not yet available to develop large-scale plantations.

Following Batara, the founders, experts, and professionals involved with its establishment have agreed to initiate the next phases of developing large-scale Pongamia plantations and entering the HVO production industry. At the same time, Batara has initiated efforts to introduce planting Pongamia as a solution for carbon sequestration and build collaborations with the local government institutions to establish conservation programs in Lembata. Each of the subsequent phases will be organized under different entities to ensure good governance. Furthermore, extending from Batara's seeds center to the HVO business require new governance arrangements and a different set of collaborations.

4.10 Discussions

Based on the analyses of the Styrax case and Pongamia case, the common trait identified is the complex sociocultural and socioeconomic factors that are intertwined with the business activities, which involves planting and farming activities. Businesses source land and human resources from the location where the organization is set up. Daily business activities will have direct interactions with the communities, and therefore, social context is a critical consideration in business model design. For the Styrax case, the business model represents current practices based on sociocultural inheritances and socioeconomic concerns. Similarly, the decision to establish Batara in Lembata in the Pongamia case is driven by social ties and availability of natural Pongamia trees throughout the island.

Another common trait is the entangled value chain from upstream, which includes planting and harvesting, to downstream, which consists of processing, trading, and manufacturing of bio-based products. While analyzing the business model generally involves examining one organization, bio-based business model requires a complete overview of the entire chain due to the interrelated connection between activities. For the Styrax case, quality and quantity of Styrax benzoin produced depends on the quality and quantity of Styrax, which is mainly determined by how well the benzoin gardens are managed. The Pongamia case also demonstrated the significance of quality seeds and seedlings to have a productive HVO business. Any development or improvement in the middle or downstream needs to consider the circumstances related with the upstream.

In addition to the commonalities, this study found major differences between the Styrax and the Pongamia cases. Analysis found that current practices of cultivating Styrax benzoin remain unchanged, which indicates the pressing need for developing an innovative business model and building agility. While the significance of Styrax benzoin as a bio-ingredient in the global market has grown, attention towards improving productivity appears to be less pronounced. Cultivation of Styrax in North Sumatra adheres to the ways of the ancestors without organized attempts to improve performance so as to produce better quality at stable or increasing quantity. Although recognized to be the main source of income, the benzoin gardens are managed in their present conditions and with the absence of exploring new cultivation practices. Instead, the Styrax communities turn to cultivating other commodities as additional sources of income.

In the case study reported by Fernandez (2004), the author provided a detailed discussion on the fundamental governance element of land ownership status. Prior to Indonesia's independence, the communal land management systems naturally formed in accordance to the needs of the land natives. When the national land reform was in place after independence, all forests and resources were determined to be under the authority of the GOI but without acknowledging those areas that are under customary tenurial systems (termed "*tanah ulayat*" or "*tanah adat*"), such as the forests where the benzoin gardens are located. With the support of the local government, these customary tenurial systems were recognized and converted into "benzoin garden rights" or the rights for use of the land consisting Styrax, including selling or renting out those land, in accordance with traditional laws (or "*hukum adat*"). To this day, ownership of those forest land is still under the GOI with benzoin garden rights given to the designated tribe and managed by the leaders. Moreover, the Styrax benzoin business is not managed under a legal entity or organized under a cooperative form but rather relies on personal or family relations. This circumstance is fragile due to the heavy reliance on key individuals' capabilities to initiate efforts for optimizing production. The declining interest in preserving Styrax benzoin practices, especially for the younger generations, further exacerbated the fragility of this condition.

Different from the Styrax case, the Pongamia case study provides insights for developing a bio-commodity business from the corporate perspective and for a professionally managed legal entity. Although still in development stage, the Pongamia case offers an alternative solution with the cultivation of Pongamia seeds as the backbone for the bio-energy business model. Sociocultural ties to Pongamia for the Lembata people are strong as shown by the abundant indigenous trees and the traditional use of Pongamia for herbal remedies or timber. However, the use of Pongamia as feedstock for bio-energy and carbon sequestration is a novel idea to the people of Lembata. Therefore, Batara's biggest challenge is ensuring social acceptance of this new business, and motivating collaborations with the local government as well as social leaders to work towards the same goals. In other words, Bataras need to ensure that its business activities coincide with the social values and traditional laws, while at the same time consistently create value.

Overall, embedding sustainability principles in formulating bio-based business models is pertinent for ensuring agility. Building agility requires developing the ability to make swift adjustments to the business model components when required. Derived from the Styrax and Pongamia cases, building agility necessitates three prerequisites for ensuring appropriate formulation and execution of business models that would lead to performance. First is building human capital through engagement and talent development. In the Styrax case, the business model was passed down from one generation to the next but remained unevolved due to the degrading interest to preserve the business as the prestige for cultivating Styrax diminished over time. In the Pongamia case, Batara is in the process of engaging the communities through social engagement programs and routine dialogues. Building engagement is necessary so that the individuals are aware of the significance of the commodity and interested to participate in working towards improving competitiveness. At the same time, fostering engagement through education and training can attract and develop talents for innovating bio-based business models. Both the Styrax and Pongamia cases showed the need for individuals with the knowledge and know-how for developing bio-based businesses, such as in cultivating quality seeds and maintaining the trees to ensure productivity.

The second prerequisite is setting up the appropriate governance of the business models. Related to the first prerequisite concerning human capital, organization and legal entity are essential for developing business model innovations. This is especially apparent in the Styrax case where the traditionalists are not operating under a legal entity but rather based on personal relations, which weakens the positioning in terms of ownership status of the benzoin gardens and negotiating with corporate buyers. Similarly, in the Pongamia case, the business model innovation designed for an integrated upstream and downstream HVO business requires critical decisions on arranging the legal entities, shareholders, collaborators, consultants and investment partners. In the business model agility lens, the third superblock is the governance and management role, which includes legal considerations, leadership arrangement and organization structure. In formulating and executing bio-based business models, the third superblock, governance and management role should be at the highest priority as the other two superblocks depend on the governance setup.

The third prerequisite is incorporating sustainability principles in the business model design. Considering the need for a complete overview of the upstream to downstream chain for each particular commodity, adopting sustainability allows for identifying innovation opportunities. For example, one of the key issues for the Styrax case is the economic returns from benzoin that fluctuates with market prices and quality. The agroforestry model of Styrax can optimize productivity of the forest areas by cultivating other commodities that will not only improve socioeconomic conditions but also help conserve the forests. In the Pongamia case, sustainability elements are already embedded in the business model design. Selection of Pongamia as the main feedstock of bio-fuel is aimed at reducing global carbon emissions from two ends, namely, sequestering carbon in the trees and making good quality HVO more accessible to replace fossil fuels. Since bio-commodities are cultivated in nature and an integral part of the social communities, a business model embedded with sustainability will build agility so as to allow for evolution of the bio-based business models.

5 Conclusion

Using the Business Model Effectiveness lens of the agility framework, the analysis results provide clarity on innovations to be initiated in the formulation and execution of the bio-based businesses. More importantly, the case studies presented in this article provided a complete picture of the realities concerning innovations in the commodity businesses in Indonesia that revolves around the delicate intricacies related to sociocultural and socioeconomic contexts. The agility analyses resulted in the identified prerequisites for the formulation and execution of bio-based business model innovations, namely, human capital, governance, and sustainability principles. These prerequisites are the underlying foundation of the bio-based business model construction. Furthermore, a bio-based business model requires an integrated overview of the entire commodity chain that will lead to innovations impacting the entire industry. This study offers a solution for designing a bio-based business model so as to build the agility necessary for the business to appropriately evolve.

Future studies on other bio-commodities using the agility framework are necessary to develop business model standards aimed at improving quality, increasing quantity, and overall maximizing value. This article marks the beginning of further exploration on the subject of bio-based business models and can enlighten the path for exploration of novel and better ways to innovate and create sustainable value. The realities discovered during this study provides a glimpse of actual business practices that are prevalent in Indonesia. In addition, this article offers an explanation of the different roles involved in formulating an innovative bio-based business model that includes traditional communities, practitioners, government institutions, and non-government organizations. Increasing awareness of the potential solutions may lead to building common interests and goals of betterment. Ultimately, this article intends to motivate collaborations and commitment for innovating the bio-based business models to ensure their sustainability and value creation for all stakeholders.

Acknowledgements This paper is an extension of previous studies on agility and organized by the Center for Innovation Excellence of PT Kresna Prakasa Utama. The authors would like to thank PT Lembata Hira Sejahtera (Batara) for participating in this study.

References

Aagaard, A., & Lindgren, P. (2015). The opportunities and challenges of persuasive technology in creating sustainable innovation and business model innovation. *Wireless Personal Communications, 81*(4), 1511–1529.

Amit, R., & Zott, C. (2012). Creating value through business model innovation. *MIT Sloan Management Review*. https://sloanreview.mit.edu/article/creating-value-through-business-model-innovation/.

Balasooriya, D., Karunarathna, C., & Uluwaduge, I. (2021). Wound healing potential of bark paste of Pongamia pinnata along with hirudotherapy: A case report. *Journal of Ayurveda and Integrative Medicine, 12*(2), 384–388. https://doi.org/10.1016/j.jaim.2021.01.014

Baumeister, C., Scherer, A., & Wangenheim, F. V. (2015). Branding access offers: The importance of product brands, ownership status, and spillover effects to parent brands. *Journal of the Academy of Marketing Science, 43*(5), 1–15.

BPS. (2023). Luas Tanaman dan Produksi Kemenyan Tanaman Perkebunan Rakyat menurut Kabupaten/Kota. https://sumut.bps.go.id/indicator/54/211/1/luas-tanaman-dan-produksi-kemenyan-tanaman-perkebunan-rakyat-menurut-kabupaten-kota.html.

Chesbrough, H. (2010). Business model innovation: Opportunities and barriers. *Long Range Planning, 43*(2–3), 354–363. https://doi.org/10.1016/j.lrp.2009.07.010

Degani, E., Prasad, M. V. R., Paradkar, A., Pena, R., Soltangheisi, A., Ullah, I., Warr, B., & Tibbett, M. (2022). A critical review of Pongamia pinnata multiple applications: From land remediation and carbon sequestration to socioeconomic benefits. In *Journal of environmental management* (Vol. 324). Academic Press. https://doi.org/10.1016/j.jenvman.2022.116297.

Fernandez, C. G. (2004). Benzoin, a Resin Produced by Styrax Trees in North Sumatra Province, Indonesia. In K. Kusters & B. Belcher (Eds.), *Forest products, livelihoods and conservation: Case studies of non-timber forest product systems* (Vol. 1). Center for International Forestry Research (CIFOR).

Fontana, A., & Zubaedah, S. Y. (2012). A conceptual model on the relationships between business strategy, business model innovation, resource configuration and performance. *Jurnal Manajemen Teknologi, 11*(2), 167–194. https://scholar.google.com/scholar?oi=bibs&cluster=13729962123388440968&btnI=1&hl=id.

Harada, K., Wiyono, W., & Munthe, L. (2022). Production and commercialization of benzoin resin: Exploring the value of benzoin resin for local livelihoods in North Sumatra, Indonesia. *Trees, Forests and People, 7*. https://doi.org/10.1016/j.tfp.2021.100174.

He, Q., Sun, Y., Chen, X., Feng, J., & Liu, Y. (2023). Benzoin resin: An overview on its production process, phytochemistry, traditional use and quality control. *Plants, 12*(1976). https://doi.org/10.3390/plants12101976.

Hossain, M. (2017). Business model innovation: Past research, current debates, and future directions. *Journal of Strategy and Management, 10*(3), 342–359. Emerald Group Publishing Ltd. https://doi.org/10.1108/JSMA-01-2016-0002.

Kiswandono, A. A., Heri Iswanto, A., Susilowati, A., & Lumbantobing, A. F. (2016). Analysis of the content of cinnamic acid and phytochemical screening sap incense type of Bulu (Styrax Benzoine Var. Hiliferum) of North Tapanuli. In *Prosiding Seminar Nasional Kimia-Lombok*.

Miles, R. E., & Snow, C. C. (2003). *Organizational strategy, structure, and process*. Stanford University Press.

Oils and Fats International. (2018). HVO making it big. In *Oils and fats international magazine* (pp. 18–20). https://www.ofimagazine.com/content-images/news/HVO2.pdf.

Reeves, M., Levin, S., & Ueda, D. (2016). The biology of corporate survival: Natural ecosystems hold surprising lessons for business. *Harvard Business Review*. https://hbr.org/2016/01/the-biology-of-corporate-survival.

Schaltegger, S., & Burritt, R. (2005). Corporate sustainability. In H. Folmer & T. Tietenberg (Eds.), *International yearbook of environmental and resource economics 2005/2006* (pp. 185–222).

Schaltegger, S., Hansen, E. G., & Lüdeke-Freund, F. (2016). Business models for sustainability: Origins, present research, and future avenues. *Organization and Environment, 29*(1), 3–10. SAGE Publications Inc. https://doi.org/10.1177/1086026615599806.

Senge, P. M., & Carstedt, G. (2001). Innovating our way to the next industrial revolution. *MIT Sloan Management Review, 42*(2), 24–38.

Shafer, S. M., Smith, H. J., & Linder, J. C. (2005). The power of business models. *Business Horizons, 48*(3), 199–207. https://doi.org/10.1016/j.bushor.2004.10.014

Stubbs, W., & Cocklin, C. (2008). Conceptualizing a "sustainability business model." *Organization and Environment, 21*(2), 103–127. https://doi.org/10.1177/1086026608318042

Wade, M., Joshi, A., & Teracino, E. A. (2021). 6 principles to build company's strategic agility. *Harvard Business Review*. https://hbr.org/2021/09/6-principles-to-build-your-companys-strategic-agility.

World Atlas. (2023). *The 10 most biodiverse countries in the world*. Worldatlas.Com. https://www.worldatlas.com/nature/the-10-most-biodiverse-countries-in-the-world.html

Zubaedah, S. Y. (2016). Building agility: Firm configuration for innovation and growth (A. Fontana, Ed.; First). Penerbit Universitas Indonesia (UI Press).

Zubaedah, S. Y. (2022). Reading the signs: Incorporating knowledge creation within strategy formation to attain performance. In *International Conference on Business and Management Research* (pp. 88–97). https://www.atlantis-press.com/proceedings/icbmr-20.

Zubaedah, S. Y., & Fontana, A. (2023). The three-lens approach to agility: Capturing distinct strategic views on the organization for comprehensive examination of firm agility. *Journal of Economics, Finance and Management Studies, 06*(09), 4202–4218. https://doi.org/10.47191/jefms/v6-i9-08.

Vortex Bladeless Wind Turbines for Airport Runway Energy Generation: Conceptual and Economic Feasibility Studies

Hasan Tariq Hamdan, Ahmad Al Ramahi and Sharul Sham Dol

Abstract

The global demand for sustainable energy is growing rapidly, especially in the aviation sector. The target of this project is to design an environmentally friendly vortex bladeless wind turbine (VBWT) to be applied at the airport runaway. The project largely centered on the extensive conception of the vortex technology, energy endurance, and the vortex bladeless turbine design using computational fluid dynamics simulations (CFD) of viscous shear–stress transport SST k-ω turbulence model (ANSYS FLUENT). Vortex-induced vibration (VIV) is the result of the classical Kármán vortices generated by alternating vortex formation due to the instability of the weakening boundary layers by adverse pressure gradient. The oscillatory flow phenomenon can typically be seen in the flow passing through a cylindrical or bluff-body-shaped object. The fluctuation depends on the unsteady lift and drag forces generated and the effects become greater for free-end three-dimensional flexible cylinders. This mechanical oscillation dynamics can be enhanced via resonance and if it could be harnessed appropriately, would be converted into useful energy for airport fundamental operations.

Keywords

Renewable energy · Aerodynamic analysis · Environmental impact assessment · LCOE · VIV · CFD · Kármán Vortices

H. T. Hamdan · A. Al Ramahi · S. S. Dol (✉)
Department of Mechanical and Industrial Engineering, Abu Dhabi University, P.O. Box 59911, Abu Dhabi, United Arab Emirates
e-mail: sharulshambin.dol@adu.ac.ae

H. T. Hamdan
e-mail: 1065775@alumni.adu.ac.ae

A. Al Ramahi
e-mail: ahmad.alramahi@adu.ac.ae

1 Introduction

Vortex bladeless wind turbines can be applied at airport runways, in which vortices are generated from the airplane fuselages and wings that depart and arrive via the runway during take-off and landing maneuvers (Fig. 1). These fluctuations forces (i.e., lift and drag) can be harnessed to generate power by the vortex instabilities at large Reynolds number mainly due to the aircraft size (Eq. 1).

$$Re = \frac{\rho v_t D}{\mu} \quad (1)$$

Vortex-induced vibration is the result of the classical Kármán vortices generated by alternating vortex formation due to the instability of the weakening boundary layers by adverse pressure gradient (Fig. 2) (Dol et al., 2021). The oscillatory flow phenomenon can typically be seen in the flow passing through a cylindrical or bluff-body-shaped object. The fluctuation depends on the unsteady lift and drag forces generated and the effects become greater for free-end three-dimensional cylinder (Yong et al., 2017). This mechanical oscillation dynamics can be enhanced via synchronization through resonance and later is converted to electrical energy (Eq. 2).

$$P = v_t F_D \sin(2\pi F_N t) \quad (2)$$

There are no moving components, such as gears, bearings, or the requirement for maintenance, in a VBWT. Compared to conventional turbines, it is less noisy and requires less maintenance, it is dependent on the aero-elastic resonance phenomenon, by directly harnessing the energy of the wind from the passing airplanes (Abdullah et al., 2020). VBWT design has a number of attractive features; being self-running and independent, made of lightweight minimal raw materials (i.e., aluminium alloy), environmentally friendly, in addition to its low cost and harmlessness to the airport operations. The performance depends on the following:

M. Amer et al. (eds.), *Territorial Landscapes Solutions for Sustainable Energy and Food Security*, Sustainable Landscape Planning and Natural Resources Management, https://doi.org/10.1007/978-3-031-82419-7_5

Fig. 1 Wing-tip vortices generated during landing and taking-off operations

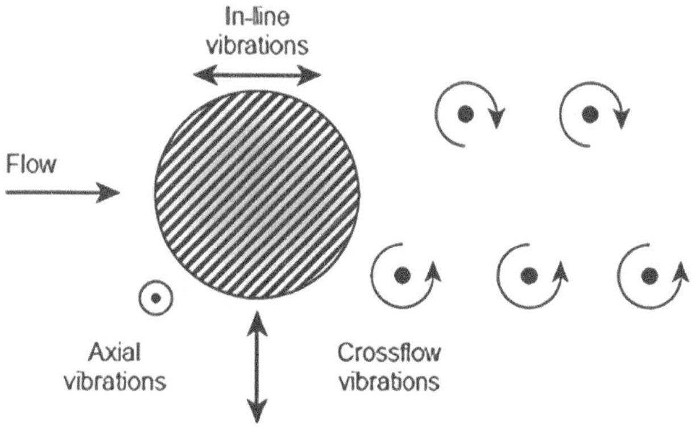

Fig. 2 Vortex-induced vibration

- Frequency and speed of taking-off and landing of airplanes.
- Size of airplanes (i.e., fuselage and wingspan).
- Air temperatures and pressures.
- Size and materials of VBWT.

2 Conceptual Study

Airport runways are meant to operate on a large frequency of air traffic on a constant basis. Large aircrafts produce fluctuating turbulence effects from the landing and take-off maneuvers. These procedures are repeated with specific frequency variations based on the size and length of the runways, the number of departures and arrivals, and the size of the airplanes with the normally large take-off and landing speeds. Densely populated metropolitans tend to have larger airports and having more volume of flights taxing the runways with larger-size aircrafts (i.e., Dubai, New York, Toronto, London, Singapore, Kuala Lumpur, and Tokyo). When airplanes arrive and depart, eddying turbulent flow will be generated behind the aircraft and its occurrence will stay for some period before getting dissipated. These fluctuations, if harnessed appropriately, would be converted into useful energy for airport fundamental operations (i.e., lighting, flow pump refueling activities, heating, and ventilations at the airport buildings).

Unstable flow separation (i.e., wake region) generated from airplanes depends on three parameters, including the load, wingspan, and landing and take-off speeds. For instance, statistically, Mumbai International Airport in India (i.e., Chhatrapati Shivaji Maharaj International Airport) has one of the higher numbers of aircrafts that arrive and depart the runway, with roughly 1000 daily departures/arrivals and the estimated time between two airplanes is about 1.5 min. The size of upper-medium regional airplanes (i.e., Airbus A320 family) can handle about 170 passengers, have a take-off and landing speed about 300 km/h. Based on this speed and the aircraft characteristic dimension, the Reynolds number is about 15.5×10^9, which is beyond the region of full turbulence, making it appropriate for the VBWT system. For 1000 VBWT systems, they would produce about 400 MWh per year, which is about 5–10% of the medium-size airport's typical energy demand. Therefore, even with the tiny fraction of energy that could be harnessed per VBWT system (i.e., one cylinder), the scale can be expanded for larger airport operations such as in Dubai (DXB).

3 Methodology and Preliminary Results

The computational fluid dynamics simulation CFD ANSYS model was conducted using viscous shear–stress transport SST k-ω turbulence model (ANSYS FLUENT) to simulate the air path and its oscillations and how the mast design can be optimized (i.e., by adding vortex generators) to produce the best possible VBWT aerodynamics performance hence energy generation (Dol et al., 2022; Shlash & Koç, 2022). The VBWT model was then fabricated and experimentally evaluated using a wind tunnel (Fig. 3). The smoke-wire flow visualization demonstrates the flow separation behind the free-end cylinder exhibiting oscillatory flow behavior. These fluctuating flows are studied using typical take-off air speeds for aircrafts to analyze the total performance. The utilization of carefully designed-springs enables the VBWT system to oscillate in resonance with the structural natural frequency to generate sufficient power measured by piezoelectric transducers even at low air speeds (Hatshith et al., 2016; Rajamuni et al., 2019; Younis et al., 2022).

The conceptual design will be feasibility assessed for airport runway and its airport buildings application (Fig. 4). Energy resource diversity is the key for energy sustainability and moving toward the UAE 2050 vision.

4 Business Model and Economic Feasibility

The business idea (Fig. 5) revolves around the implementation of Vortex Bladeless Wind Turbines (VBWT), an innovative technology that has the potential to revolutionize the wind energy sector in the UAE. These turbines operate without rotating blades, offering several advantages over traditional wind turbines (Elsayed, 2021; Mondal et al., 2022). They eliminate noise and vibration, making them ideal for urban and residential areas. Additionally, their unique design significantly reduces maintenance costs and enhances reliability. It is believed that VBWT can seamlessly integrate into the existing renewable energy infrastructure in the UAE and contribute to the country's commitment to clean energy.

The market potential for Vortex Bladeless Wind Turbines in the UAE is significant. The government has shown strong dedication to diversifying energy sources and reducing

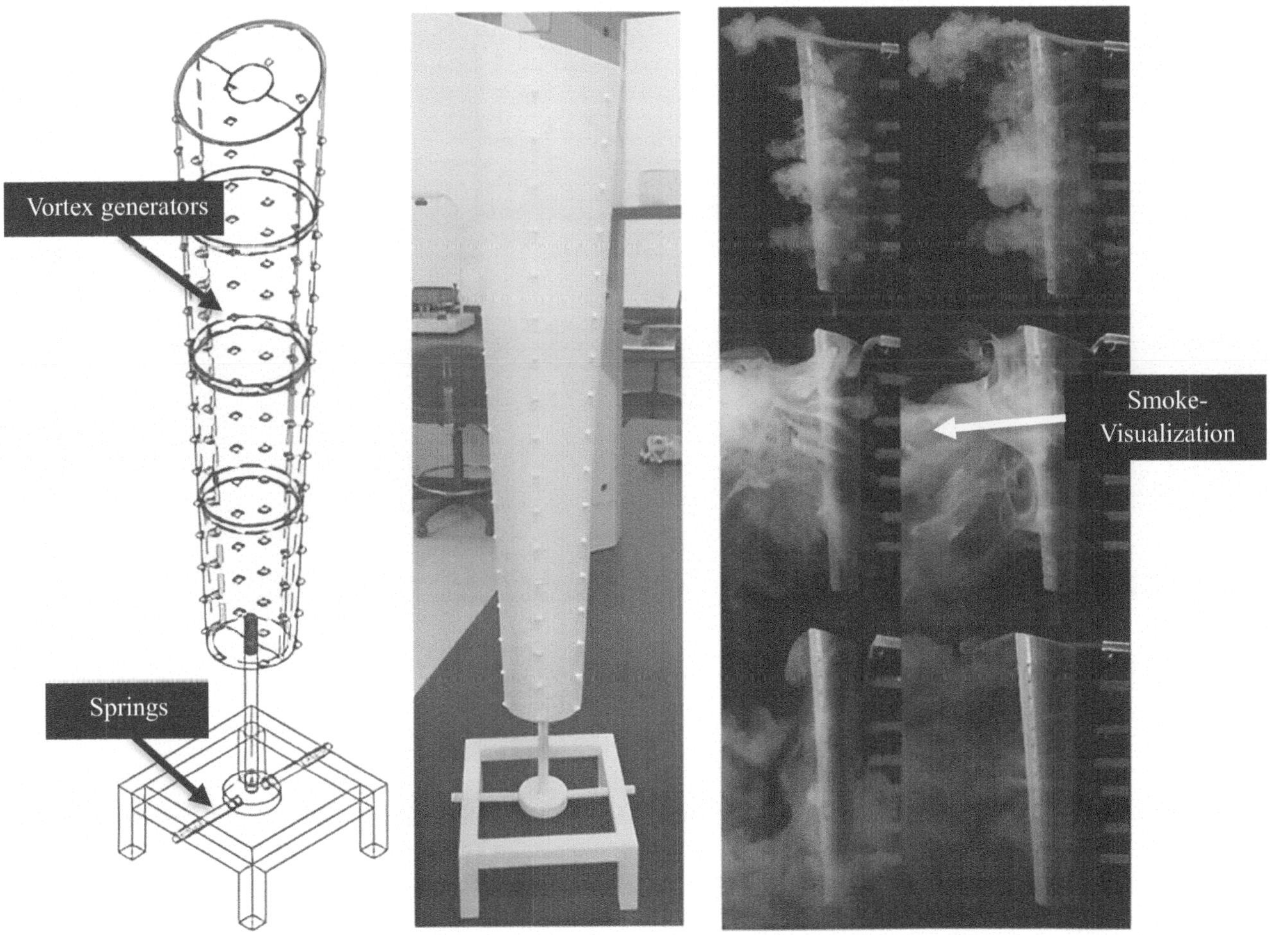

Fig. 3 VBWT model and wind tunnel studies

Fig. 4 VBWT system at airport runway

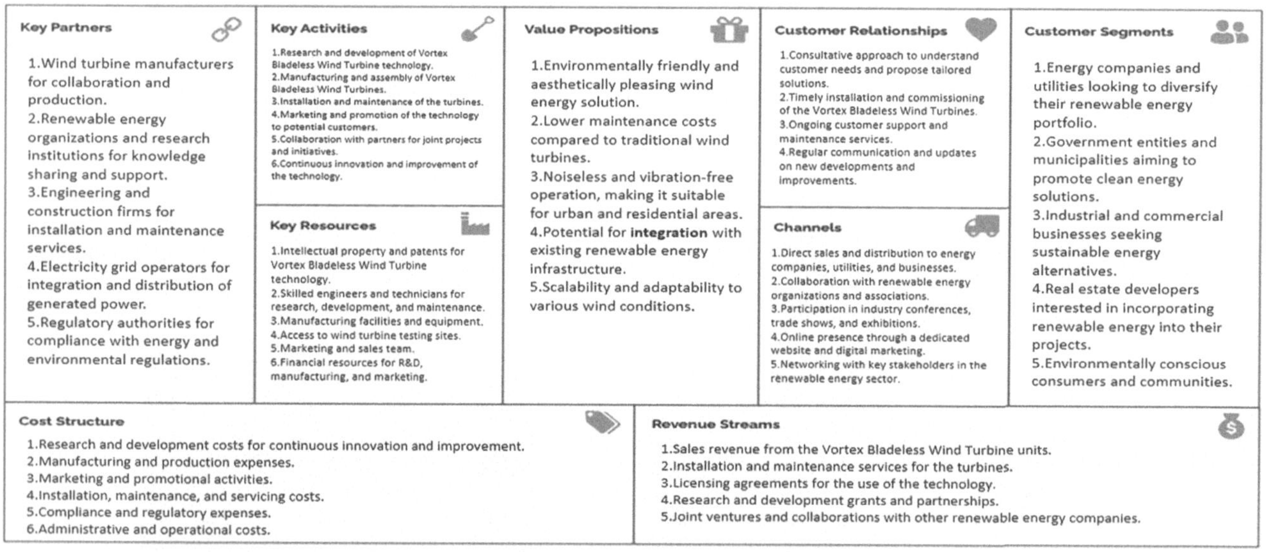

Fig. 5 VBWT business model canvas

reliance on fossil fuels. The turbines cater to a wide range of customers, including energy companies, utilities, government entities, municipalities, commercial businesses, real estate developers, and environmentally conscious consumers. The key partners have been identified, such as wind turbine manufacturers, renewable energy organizations, engineering and construction firms, and regulatory authorities, who will provide the necessary expertise, resources, and support for our initiative.

The key activities involve research and development, manufacturing and assembly, installation and maintenance, marketing and promotion, collaboration with partners, and continuous innovation. These activities will ensure the refinement of our technology, high-quality manufacturing, timely installation and commissioning, effective marketing, and ongoing improvements. In terms of financial viability, our revenue streams will come from sales of turbine units, installation and maintenance services, licensing agreements, research and development grants, and strategic joint ventures.

As per the Levelized Cost of Energy (LCOE) conducted, integrating five turbines within any location (runaways, highways, and residential buildings) would allow the product to be economically feasible as the cost of energy would reduce from 38 to 28 fils. Moreover, the more turbines are installed, the lower the LCOE becomes, and the higher energy is generated, the more profitable the product becomes (Adnan et al., 2022; Khing et al., 2015). The implementation of Vortex Bladeless Wind Turbines in the UAE presents an exceptional business opportunity. With its unique advantages, compatibility with the UAE's renewable energy goals, and the support of key partners, the business idea is poised for success.

5 Summary

By completing this project, new and important knowledge regarding the ground-breaking advancement in wind power generation would be gained that will inspire new innovative ideas for airport runaway sustainable operation. The main advantage of vortex bladeless wind turbines is that they have continuous wind vibrations from the passing airplanes, not to be affected by low natural wind speed, making theses suitable for large and busy airport applications. Bladeless turbines can have several environmental benefits on airports and the UAE, including the reduction of carbon emissions, noise pollution, and the visual impact of traditional wind turbines. Additionally, they can promote the development of renewable energy technologies and contribute to a more sustainable future.

References

Abdullah, A. Q., Abdul Azeez, A., Dol, S. S., Khan, M., Meor Said, M. A. (2020). Simulation study on vortex-induced vibration air wake energy for airport runaway application: A preliminary analysis. *Platform: A Journal of Engineering, PAJE, 4*, 38. https://doi.org/10.61762/pajevol4iss3art8420.

Adnan, N. F., Lee, K. Q., Kang, H. S., Wong, K. Y., & Tan, H. Y. (2022). Preliminary investigation on the energy harvesting of vortex-induced vibration with the use of magnet. *PROGEE, 21*, 1–7. https://doi.org/10.37934/progee.21.1.17

Dol, S. S., Yong, T. H., Chan, H. B., Wee, S. K., & Sulaiman, S. A. (2021). Turbulence characteristics of the flexible circular cylinder agitator. *Fluids, 6*, 238. https://doi.org/10.3390/fluids6070238

Dol, S. S., Khamis, A., Abdallftah, M. T., Fares, M., & Pervaiz, S. S. (2022). CFD analysis of vertical axis wind turbine with Winglets. *RERA, 3*, 51–59. https://doi.org/10.22044/rera.2021.11122.1074

Elsayed, A. M. (2021). Design optimization of diffuser augmented wind turbine. *CFDL, 13*, 45–59. https://doi.org/10.37934/cfdl.13.8.4559

Hatshith, K., Fernandes, B. S., Shreerama, P. R., & Raj, T. (2016). Bladeless wind power generation. *International Journal for Scientific Research & Development, 4*(3), 1897–1901.

Khing, T. Y., Zahari, M. A., & Dol, S. S. (2015). Application of vortex induced vibration energy generation technologies to the offshore oil and gas platform: The feasibility study. *International Journal of Aerospace and Mechanical Engineering, 9*(4), 661–666. https://doi.org/10.5281/zenodo.1096534

Mondal, M., Didane, D. H., Ali, A. H. I., & Manshoor, B. (2022). Technical assessment of wind energy potentials in Bangladesh. *ARFMTS, 96*, 10–21. https://doi.org/10.37934/arfmts.96.2.1021

Rajamuni, M. M., Thompson, M. C., & Hourigan, K. (2019). Vortex-induced vibration of elastically-mounted spheres: A comparison of the response of three degrees of freedom and one degree of freedom systems. *Journal of Fluids and Structures, 89*, 142–155. https://doi.org/10.1016/j.jfluidstructs.2019.02.005

Shlash, B. A. A., & Koç, I. (2022). Turbulent fluid flow and heat transfer enhancement using novel vortex generator. *ARFMTS, 96*, 36–52. https://doi.org/10.37934/arfmts.96.1.3652

Yong, T. H., Chan, H. B., Dol, S. S., Wee, S. K., Kumar, P. (2017). The flow dynamics behind a flexible finite cylinder as a flexible agitator. *IOP Conference Series: Materials Science and Engineering, 206*, 012033. https://doi.org/10.1088/1757-899X/206/1/012033.

Younis, A., Dong, Z., ElBadawy, M., AlAnazi, A., Salem, H., & AlAwadhi, A. (2022). Design and development of bladeless vibration-based piezoelectric energy-harvesting wind turbine. *Applied Sciences, 12*, 7769. https://doi.org/10.3390/app12157769

Black Sea Blue Transitions: Multi-actor Forums to Promote a Sustainable Blue Economy

Ebun Akinsete, Lydia Papadaki and Phoebe Koundouri

Abstract

The Blue Economy, which incorporates all economic activities related to oceans, seas, and coastal areas, is vital to the sustainable development of regions across the globe. The Black Sea, located at the crossroads of Europe and Asia, has enormous potential for the growth of a prosperous Blue Economy. This potential is, however, accompanied by a unique set of challenges that must be addressed to assure the sustainable growth of maritime industries in the region. These problems are addressed by DOORS Black Sea (https://www.doors-blacksea.eu), an EU-funded project that develops optimal and open Black Sea research support. DOORS connects residents, research, and industry to regenerate the Black Sea and create "blue economy" prospects by establishing a system of systems (SoS) to address human and climate change impacts on the marine ecosystem. DOORS' success, value, and effect depend on stakeholder engagement. They progress science and technology with researchers, making project work more meaningful. Multi-actor Forums (MAFs) bring together national stakeholders from Romania, Bulgaria, Turkey, and Georgia of all backgrounds to help scientists prioritize Black Sea issues with a focus on blue economy policies and innovations to fill gaps. This method also helps co-design the region's System of Systems to give researchers the datasets they need to solve environmental problems and grow the blue economy. This study analyzes how findings may affect the long-term expansion of the Blue Economy and related policy in the region.

Keywords

Multi-actor forums · Living labs · Co-creation · Blue economy · Black sea · Systems approaches

1 Introduction

Every region is affected by climate change. Sea levels are rising as polar ice barriers melt. Heat waves and droughts are increasing in some locations and extreme weather events and rainfall are increasing in others. If climate action is delayed, these effects will worsen. Climate change threatens many aspects of life. Sea levels rose throughout the twentieth century and increased recently. Ocean thermal expansion from warming is driving the rise. Melting glaciers and Antarctic ice sheets contribute. How fast the Antarctic ice sheet melts will raise Europe's sea level by 60–80 cm by the end of the century.

A third of EU residents live within 50 km of the shore, which generates 30% of GDP. Assets within 500 m of Europe's waters are worth 500–1000 billion euros (European Commission, 2023). Sea level rise, along with other climate change impacts, will exacerbate coastal flooding and erosion, affecting people, infrastructure, companies, and environment. Sea level rise will also diminish freshwater availability by pushing seawater deeper into subsurface water tables. This may also increase saltwater incursion into freshwater, impacting agriculture and drinking water. Coastal biodiversity and natural services and goods would

E. Akinsete · L. Papadaki (✉) · P. Koundouri
Sustainable Development Unit, Athena RC, Marousi, Greece
e-mail: lydia.papadaki@athenarc.gr

E. Akinsete · L. Papadaki · P. Koundouri
School of Economics and ReSEES Research Laboratory, Athens University of Economics and Business, Athens, Greece

E. Akinsete · L. Papadaki · P. Koundouri
UN SDSN Global Climate Hub, Athens, Greece

P. Koundouri
Department of Technology, Management and Economics, Denmark Technical University (DTU), Kongens Lyngby, Denmark

P. Koundouri
UN SDSN Europe, Paris, France

P. Koundouri
UN SDSN Greece, Athens, Greece

M. Amer et al. (eds.), *Territorial Landscapes Solutions for Sustainable Energy and Food Security*, Sustainable Landscape Planning and Natural Resources Management, https://doi.org/10.1007/978-3-031-82419-7_6

be affected. Many wetlands will be gone, endangering rare bird and plant species and storm surge protection.

The Blue Economy (BE) is the contribution of seas, oceans, and other marine or aquatic ecosystems to the prosperity of a country or region. With an annual added value of 500 billion EUR, the blue economy of the EU is recognized as a significant driver of economic growth in Europe and has been identified as a growth priority (European Commission, 2012). Simultaneously, European seas are overexploited and subject to multiple human pressures (Korpinen et al., 2012); this has prompted the European Commission, in accordance with the goals of the European Green Deal, to acknowledge the need to transition to a sustainable Blue Economy (European Commission, 2021). This can only be accomplished by preserving marine ecosystem services and their resistance to multiple human-caused stresses.

The Black Sea (BS) is a complex socio-ecological ecosystem that is rich in resources, and it is nested inside a dynamic geo-political region. Because of this, the Black Sea presents both basic challenges and enormous opportunity within the Blue Economy sectors. The Mediterranean Sea, Asia, and the Middle East are connected by a "strategic bridge" in southeastern Europe, on the EU's periphery. The countries' high economic potential, differing interests, and open conflict make it a vibrant, diversified region (European Commission, 2022). However, the region's Blue Economy remains underdeveloped, with huge economic potential. Key regional actors highlighted its importance for regional growth in the Burgas Vision paper and the Common Maritime Agenda (European Commission, 2018a, 2019). Despite being resource-rich, the Black Sea is one of the most polluted oceans in the world and also represents the bad environmental state of European oceans. Poor water quality has devastated fish populations and harmed species diversity. Poor environmental status has socio-economic ramifications on employment, food security, tourism, and health, requiring an urgent shift to a more sustainable Blue Economy development trajectory.

DOORS Black Sea is a project that is being funded by the EU with the intention of establishing optimal and open research support for the Black Sea in order to address these difficulties. The DOORS project aims to unite citizens, scientists, and industry for the crucial regeneration of the Black Sea, sparking a new surge of potential for the "blue economy" by developing a system of systems (SoS) to address the consequences of humans and climate change on the marine ecosystem. This will be accomplished by connecting citizens, scientists, and industry. The success, value, and impact of DOORS are all determined by the level of engagement of the many stakeholders. They do this in collaboration with researchers in order to develop scientific and technological knowledge, which in turn makes the project work more meaningful on the ground.

A widespread approach to engaging policymakers, citizens, scientists, and industry is through living labs, which aim to engage stakeholders in an active manner over an extended duration (Ogonowski et al., 2013). By employing the living labs methodology, it becomes feasible to integrate the worth of concepts and ideas that have been co-created into services that subsequently integrate into society. There is no unified definition of living labs at this time. There are numerous contexts, processes, and meanings associated with the word (Yasuoka et al., 2018). There have been scholarly investigations into living labs for more than ten years (Ballon & Schuurman, 2015; Hossain et al., 2019; Leminen & Westerlund, 2017). The extant body of literature examines living labs not only as a captivating subject matter that offers numerous research prospects for forward-thinking scholars but also as an innovative approach, methodology, and design for practitioners to surmount a myriad of contemporary challenges and necessities (Rodrigues & Franco, 2018; Voytenko et al., 2016). At present, a considerable proliferation of operational living labs can be observed, with Europe hosting the highest concentration, followed by Asia and America, and limited research conducted in the Black Sea region. To effectively assess the BE policy framework in the Black Sea, it is critical to possess a comprehensive understanding of the stakeholders' needs and priorities and subsequently adapt it to novel policy recommendations.

Multi-actor Forums (MAFs), which is essentially Living Labs, are being implemented here as a different form of stakeholder engagement structure. These forums are bringing together national stakeholders from Romania, Bulgaria, Turkey, and Georgia of all different backgrounds to assist scientists in prioritizing Black Sea issues with a focus on blue economy sectors and policies and the use of innovations to fill identified gaps. This strategy also contributes to the co-design of the System of Systems for the region, with the goal of providing researchers with the datasets they need to resolve environmental concerns and establish strategies for the growth of the blue economy in the region. The System of Systems is designed with the intention of giving researchers with the datasets they need to do their jobs effectively. This study explores the prospective consequences that established and emergent blue economy sectors hold for the long-term sustainable blue growth of the Black Sea and related policy in the region.

2 Methodology

The System Innovation Approach (SIA) allows systemic transformation through an interconnected collection of inventions that impact each other and innovate both in the system and its pieces (De Vicente Lopez, 2016). This research uses transition management and system thinking to solve persistent problems and promote sustainability (Loorbach et al., 2007; Meadows, 2008; Roorda et al., 2014). Instead of focusing on single functions or individual/sectoral advantages, the cross-sectoral system "as a whole" and its diverse actors are emphasized. Stakeholders actively participate in LLs to co-identify a linked collection of innovations to drive the desired change. LLs are open innovation venues that encourage user co-creation to better meet stakeholder demands. In an iterative approach, stakeholders from many areas and scales solve issues, co-produce knowledge, and co-design solutions (Geels & Schot, 2007). They map significant difficulties and needs related to the Black Sea Blue Economy and co-identify common aims in a desirable future vision. Creating a sustainable future requires imagining ideal possibilities, which provides direction for positive change (Bennett et al., 2021; Milkoreit, 2017; Riedy & Waddock, 2022). Pathways will be then co-developed to push the region toward a sustainable state and achieve national goals.

As an alternative method of facilitating stakeholder interaction, multi-actor forums, often known as MAFs, are now being established here. These forums are bringing together national stakeholders from Romania, Bulgaria, Turkey, and Georgia of all different backgrounds to assist scientists in prioritizing Black Sea issues with an emphasis on blue economy policies and the application of innovations to fill identified gaps. These forums are being held in order to assist scientists in prioritizing Black Sea issues.

The first step in conducting the MAFs in each of the BS countries is stakeholder mapping. Stakeholders were mapped and analyzed with the aid of an "influence/interest" matrix (Fig. 1), whereby "influence" refers to how much power and capacity the stakeholder has to effect change, and "interest" refers to how likely the stakeholder is to engage in activities or initiatives relevant to the case study focus; this may be due to resultant benefit or adverse impact (Eden & Ackermann, 1998). Once the plotting is completed, the map is validated by an external expert (e.g., project advisory board members, or other local experts).

By utilizing this matrix, we were capable of evaluating the position of every stakeholder in relation to these two criteria. This analysis aimed to ascertain which subset of stakeholders is most suitable for participation in the MAFs. The stakeholders located in the upper right quadrant, which represents those with significant influence and interest, will comprise the core group from which the MAF members will be chosen (see Table 1). Nevertheless, the inclusion of stakeholders from the lower right and upper left quadrants (i.e., those with low interest and low influence) is also taken into account, as the former group comprises individuals who may be instrumental in the implementation of potential recommendations and the latter group comprises

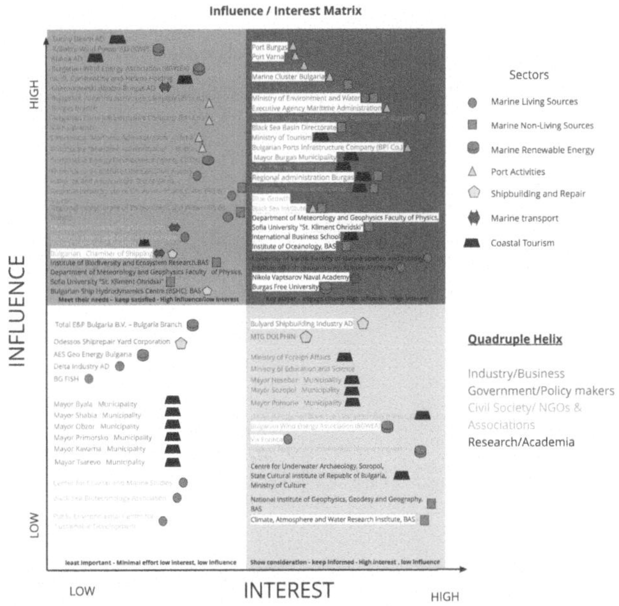

Fig. 1 Influence–interest matrix (example from MAF in Bulgaria)

individuals who possess extensive local expertise but generally lack decision-making authority (i.e., their opinions are often overlooked). By employing this methodology for participant selection in the MAF, it becomes possible to discern the stakeholders who are most pertinent to the study and are more inclined to be involved in the research endeavors. Although there are benefits to using MAFs as a platform to involve influential decision-makers, identifying stakeholders who are willing to dedicate time and effort to aiding the research process is also possible through the examination of "Interest" (Brugha & Varvasovszky, 2000; Mendelow, 1981).

The first round of MAFs took place between November 2022 and February 2023 in Romania, Bulgaria, Turkey, and Georgia and was conducted in local language. The primary aim of these workshops was to generate a comprehensive map of the regional blue economy and establish a prioritized inventory of requirements pertaining to the local blue economy. The participants were provided with an overview of the DOORS project, including its objectives and those of the MAF, throughout the workshop. The MAF undertook a regional context mapping of their blue economy systems concerning Black Sea BE sectors (see Fig. 3), as defined in the 2018 annual economic report on the EU blue economy (see Fig. 2) (European Commission, 2018b). In the context of BE development, MAF members were then engaged in a facilitated discussion to assess and prioritize these needs using the PESTLE approach (see Fig. 4) to validate the results of these deliverables. The objective of this exercise was to enhance our comprehension of their requirements and policy deficiencies by integrating their knowledge and

Table 1 Levels of stakeholders' engagement based on influence and interest

Influence/Interest relationship	Level of engagement	DOORS MAF engagement
High interest/High influence	Manage closely (High involvement, tap into stakeholders' expertise)	MAFs Data collection (interviews) Diffusion of project outputs
High interest/Low influence	Keep informed (one-way communication, presentation of research results)	MAFs Diffusion of project outputs
High influence/Low interest	Keep satisfied (Acting together with stakeholders based on shared goals)	MAFs Diffusion of project outputs
Low influence/Low interest	Monitor (one-way communication with minimum effort)	Diffusion of project outputs

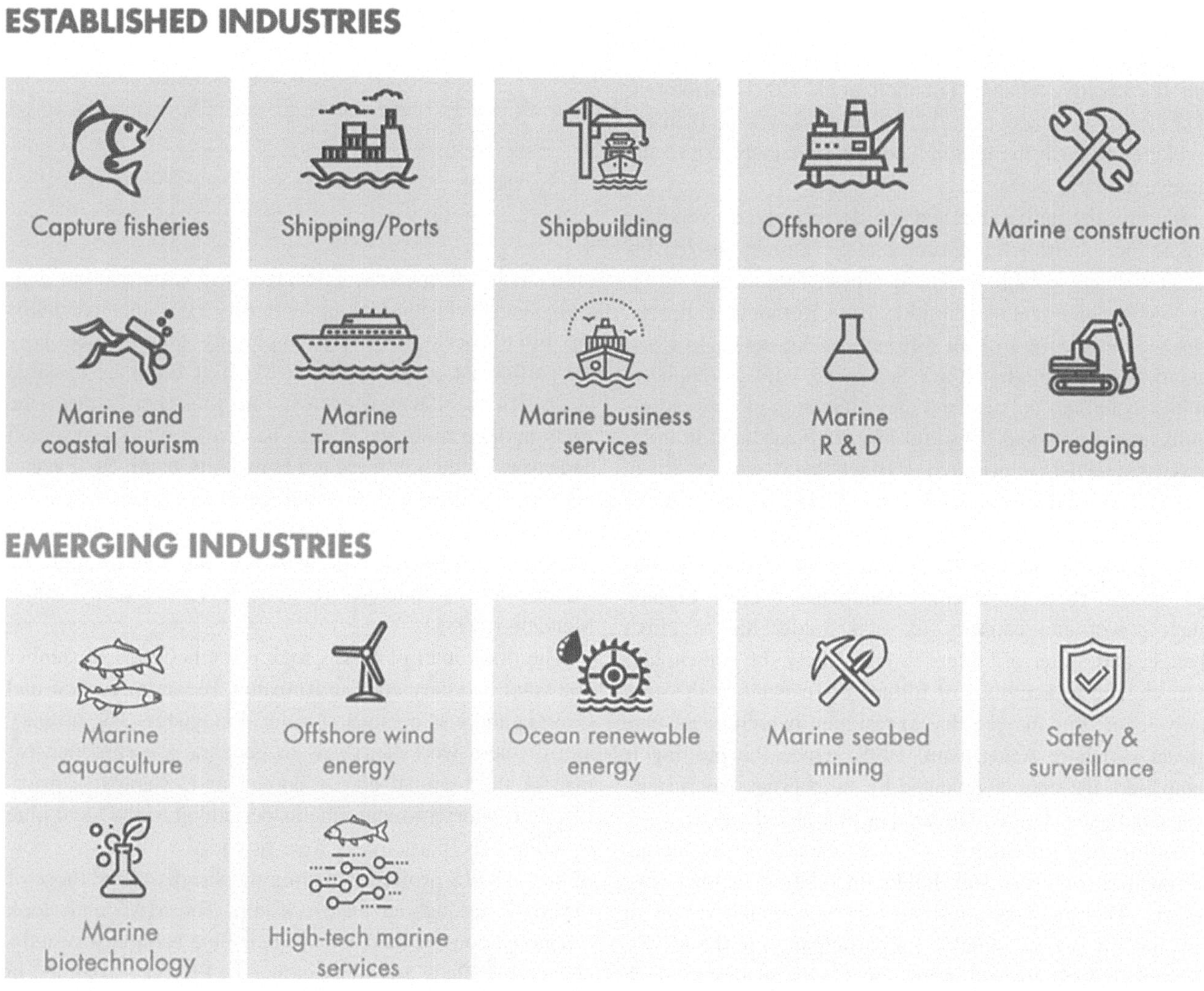

Fig. 2 Blue economy "Established" and "Emerging" industries (Caribbean Development Bank, 2018)

perspectives, as well as to incorporate any absent elements into our analysis.

To pursue uniformity across the four MAFs, facilitators, were invited to a 2-h "Train the Facilitator" workshop in October 2022, during which they were briefed on the primary objectives of the MAFs and the instruments that would be utilized during the inaugural workshop in their respective countries. In addition, they received instruction on how to facilitate the discussion and behave in a variety of potential situations.

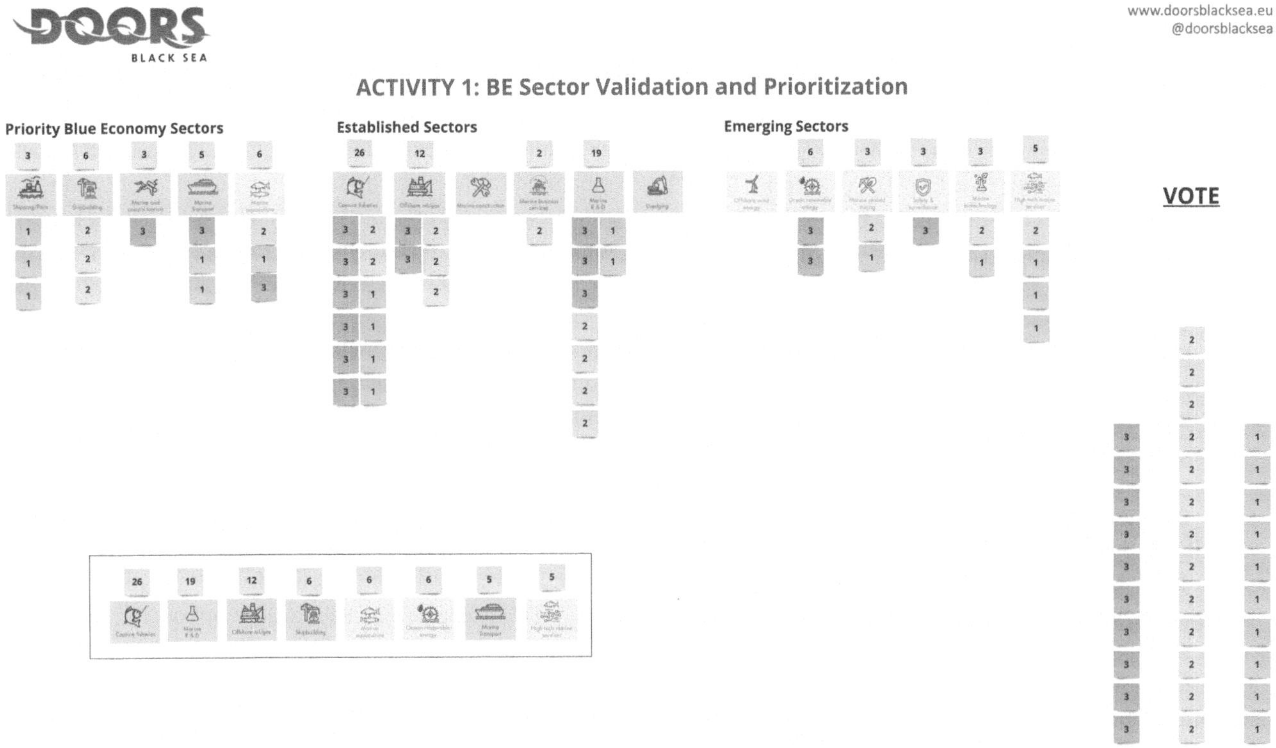

Fig. 3 First round of MAFs—blue economy sector validation and prioritization (example from MAF in Turkey)

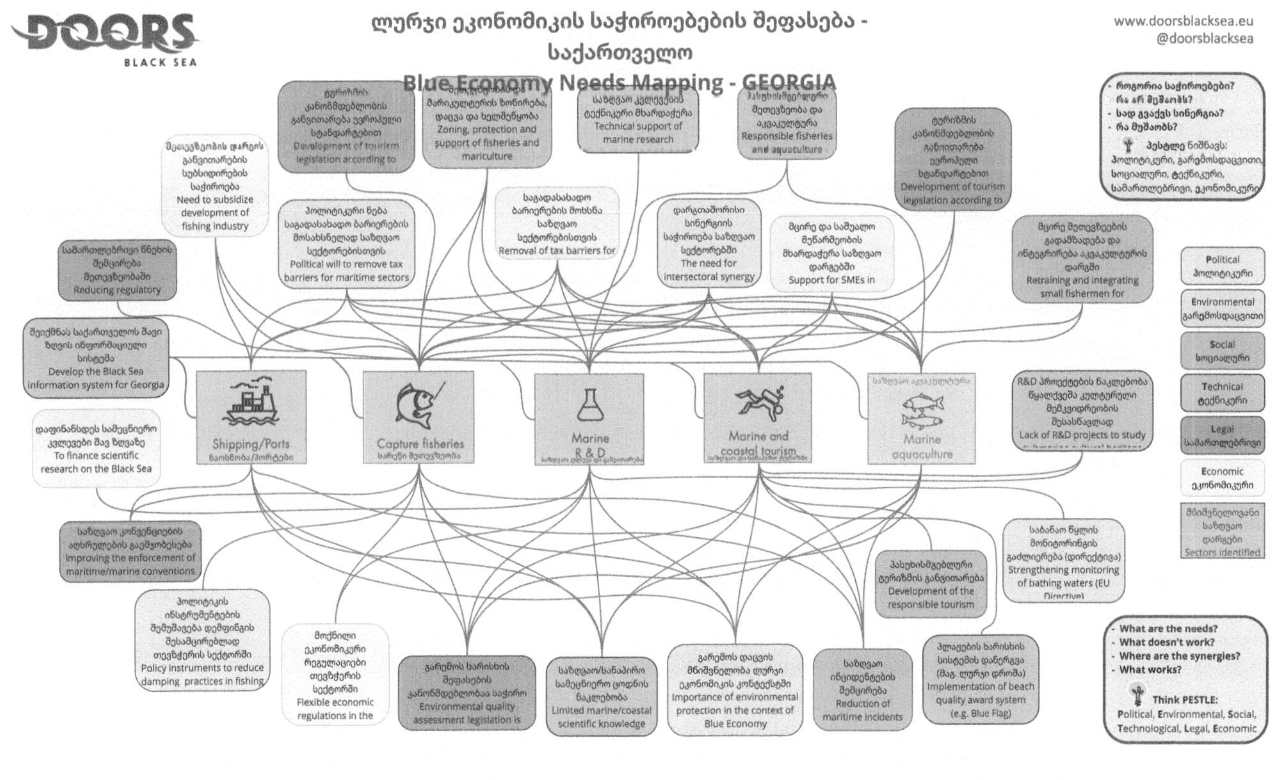

Fig. 4 First round of MAFs—blue economy needs mapping (example from MAF in Georgia)

3 Results and Discussion

The blue economy incorporates economic activities that occur directly in the ocean and seas, utilize outputs from the sea for consumption or as a source of income, or involve the ocean and seas themselves (see Fig. 2). As seen in Table 2, when it comes to the sectors that should be prioritized, stakeholders in all four countries strongly believe that the Fishing and Marine Research and Development sectors should be supported. The Black Sea region comprises 37.4% of FAO fishery area. Both Bulgaria and Romania maintain fishing interests in the Black Sea. The sole source of all landings made by the fishing fleets of Bulgaria and Romania is the Black Sea. In 2019, revenue and GVA were estimated at €10.5 million and €7.4 million, respectively (European Commission, 2022).

The Black Sea is experiencing a concurrent surge in both the popularity of tourism and the sustainability challenges that plague the sector (e.g., environmental pressures, limited area accessibility during peak visits, inadequate overall visibility). A new initiative, Tourism 4.0 for the Black Sea initiative, co-financed by the European Maritime and Fisheries Fund (EMFF), aims to empower local stakeholders from both the public and private sectors of the tourism industry to enhance their comprehension of prevailing trends, visitor impact, and flow patterns. Marine & Coastal Tourism is a significant value generation sector in Bulgaria, Georgia and Romania. Marine aquaculture seems to gain more and more field in Georgia, Turkey, and Romania. A General Fisheries Commission for the Mediterranean regional study reveals that aquaculture productivity in the Black Sea region has increased consistently over the past few years, from more than 500,000 tons of farmed seafood (primarily salmonoids, carp, and European seabass) in 2017 to more than 700,000 tons in 2019 (FAO, 2023).

Turkey and Romania have identified Ocean Renewable Energy as a key sector in their economy. Indeed, the commercial viability of floating offshore wind in challenging environments and deep waters is increasingly being achieved. Floating offshore wind for depths ranging from 50 to 1000 m appears to be a viable option for EU member states and regions lacking shallower waters; it could potentially access untapped markets in the Mediterranean Sea, the Atlantic Ocean, and the Black Sea (European

Table 2 Black Sea blue economy sectors prioritization

Sector	Definition	Established or emerging	MAFs in which this sector was prioritized
Capture fisheries	The practice of obtaining naturally occurring living resources in both freshwater and marine environments in a sustainable manner	Established	Bulgaria, Georgia, Turkey, Romania
Marine R&D	The activities centered around the advancement of technology, knowledge, and capabilities pertaining to marine environments, encompassing oceans, seas, and other aqueous bodies	Established	Bulgaria, Georgia, Turkey, Romania
Marine and coastal tourism	The provision of tourism-related services in and around littoral or marine environments, which support the local community's sustainable development	Established	Bulgaria, Georgia, Romania
Marine aquaculture	The practice of aquaculture and farming with the intention of minimizing any adverse effects on the purity of air, water, and soil	Emerging	Georgia, Turkey, Romania
Ocean renewable energy	The production of pure and renewable energy from natural sources, such as wind, wave, tidal, and solar, at sea, offshore, on land, and in close proximity	Emerging	Turkey, Romania
Shipping/Ports	The operations related to maintaining a sustainable maritime transport ecosystem, encompassing terminal services and the conveyance of passengers and cargo via water	Established	Bulgaria, Georgia
Marine business services	The commercial activities that rely on water and are associated with marinas and other vessel service operations	Established	Bulgaria
Offshore oil and gas	The extraction of gas and hydrocarbons from submerged sources	Emerging	Turkey
Shipbuilding	The goods and services necessary for the construction, upkeep, restoration, and repair of vessels used for ecologically conscious maritime transportation	Established	Turkey

Commission, 2022). On the other hand, Bulgaria and Georgia have identified shipping and port sectors as a priority. Georgia, in the last 20 years (2003–2023), the transport sector contributed an average of 145 million euros to Georgia's gross domestic product (GDP), peaking at 285 million euros in the second quarter of 2023, while in Bulgaria the port and shipping activities accounted for 115 million euros in 2020 (The World Bank, 2020; Trading Economics, 2023).

Capture fisheries, Shipping/ports, Shipbuilding, Marine and Coastal Tourism, Marine aquaculture and Ocean renewable energy, prioritized in this MAF, are also selected among the 10 key sectors, in which BlueInvest will seek to provide support. BlueInvest, funded by the EU, aims to increase investment and innovation in sustainable technologies for the blue economy by assisting early-stage businesses, SMEs, and scale-ups with financial preparedness and access.

These results are also aligned with the two Black Sea flagship strategies, the Common Maritime Agenda for the Black Sea and the Black Sea Strategic Research and Innovation Agenda (Black Sea Strategic Research and Innovation Agenda, 2019; European Commission, 2019). They both capture explicitly the role of these two sectors in the Blue Economy suggesting clear actions. Indicatively, SRIA suggests enhancing research on food systems, such as fisheries, recruitment, stock assessment and sustainability, marine protected areas (MPAs), and biotechnology (such as alternative protein sources). In terms of the sectors that are poorly mentioned in these CMA and SRIA, but prioritized in the MAFs, we see that shipbuilding and offshore oil/gas are perceived as important sectors for the local economies and should be included in these strategies.

Table 3 presents the challenges in the Black Sea region, as resulted from the first round of MAFs. It is apparent that there is need for collaboration across difference disciplines, institutions and countries to support the uptake of the Blue Economy in the Black Sea. Pollution and environmental degradation seem to be the dominant challenge across all four Black Sea countries. However, on a country level, there are different priorities. In Bulgaria, the dominant challenges seem to be the lack of investments in infrastructure in coastal zones, lack of adequate infrastructure and green energy and lack of financing scientific research for the Black Sea. In Georgia, they have to deal with marine litter and waste management, lack of beach quality award system and lack of financing scientific research for the Black Sea. Finally, Romania and Turkey put marine litter and waste management and monitoring and control on top of their priorities, with Romania highlighting the need for digitalization.

4 Conclusions and Future Research

A Blue Economy is possible in the Black Sea between Europe and Asia. This promise has restrictions that must be addressed to keep marine enterprises in the region.

Table 3 Challenges in the Black Sea region as presented in the first round of MAFs

Political	Environmental	Social
• Geopolitical instability • Lack of collaboration between all state institutions • Need for regional cooperation and intersectoral synergies • Lack of international cooperation • Insufficient political will • Lack of vision and long-term planning • Lack of a Black Sea brand at local and regional level • Lack of compliance with political obligations toward the EC	• Pollution and environmental degradation • Seawater quality • Imbalance of aquaculture sustainability and overfishing • Climate change's impact on biodiversity	• Need for training and capacity building in all Blue Economy sectors (limited human resources available) • Lack of job opportunities • Public awareness on aquaculture sector • Need for connectivity through cultural and natural heritage, between neighboring countries
Technological	**Legal**	**Economic**
• Need for marine research • Need for initiatives, such as a beach quality award system (e.g., Blue Flag) • Lack of adequate infrastructure • Lack of advanced technologies (e.g., for monitoring the fishing vessel in the region) • Need for digitization and creation of a unitary database • Need for use of non-polluting technologies	• Harmonization of national and EU legislation • Specific and integrated legislations drafted clearly and concisely, with the elimination of legislative loopholes • Lack of implementation of existing regulations • Lack of carrying out adequate monitoring and control • High bureaucracy • Corruption	• Need for state aid for fisheries and aquaculture sectors • Need for removing the tax barriers for the maritime sector • Lack of quotas in the fishing industry • Financial crisis • High international competition in marine products • Lack of Investments

EU-funded DOORS project provides optimal and open Black Sea research support to address these concerns. DOORS connects residents, research, and business to restore the Black Sea and promote "blue economy" opportunities by developing a system of systems (SoS) to address human and climate change impacts on the marine ecology. This study explores how methodological and thematic findings affect regional Blue Economy and policy growth using a participatory approach, namely, the Multi-actor Forums. Before mapping needs and possibilities, the participants in this initial round of MAF workshops, held between November 2022 and February 2023 in Bulgaria, Georgia, Romania, and Turkey, identified the key Blue Economy sectors in their country, respectively, as well as the associated challenges. This discussion and decisions was based on the local context. These findings will be validated by an online survey to ascertain the opportunities and challenges that the blue economy in the Black Sea region entailed. It will examine the expansion of various Blue Economy sectors, in addition to the support provided to entrepreneurs, the obstacles they encounter, and the priorities they establish. The results of these assessments will be cross-mapped with the existing policy framework in the Black Sea region and analyzed in order to identify policy gaps between the needs of national Black Sea Blue Economy initiatives and the strategic and policy aims of the EU. In succeeding rounds of workshops, the MAFs will work to co-create Innovation Pathways while collaborating with the project's Blue Growth Accelerator, which is a hub and support mechanism for Blue economy entrepreneurs and innovators.

Acknowledgements This paper is an output of the science project DOORS. DOORS has received funding from the European Union's Horizon 2020 Framework Programme for Research and Innovation under grant agreement No 101000518.

References

Ballon, P., & Schuurman, D. (2015). Living labs: Concepts, tools and cases. *Info*, *17*(4). https://doi.org/10.1108/INFO-04-2015-0024.

Bennett, E. M., Biggs, R., Peterson, G. D., & Gordon, L. J. (2021). *Patchwork earth: Navigating pathways to just, thriving, and sustainable futures*. https://doi.org/10.1016/j.oneear.2021.01.004.

Black Sea strategic research and innovation agenda (2019). http://connect2blacksea.org/wp-content/uploads/2019/12/Black_Sea_SRIA_Final.pdf.

Brugha, R., & Varvasovszky, Z. (2000). Stakeholder analysis: A review. *Health Policy and Planning*, *15*(3), 239–246. https://doi.org/10.1093/heapol/15.3.239.

Caribbean Development Bank. (2018). Financing the blue economy: A Caribbean development opportunity. https://www.caribank.org/publications-and-resources/resource-library/thematic-papers/financing-blue-economy-caribbean-development-opportunity.

De Vicente Lopez, J., & Matti, C. (2016). *Visual toolbox for system innovation. A resource book for practitioners to map, analyse and facilitate sustainability transitions. Transitions Hub Series*. Climate-KIC.

Eden, C., & Ackermann, F. (1998). *Analysing and comparing idiographic causal maps*.

European Commission. (2012). *Coastal and maritime tourism|The European maritime spatial planning platform*. https://maritime-spatial-planning.ec.europa.eu/sector-information/coastal-and-maritime-tourism.

European Commission. (2018a). *Burgas vision paper: A blue growth initiative for research and innovation in the Black Sea*.

European Commission. (2018b). *The 2018 annual economic report on EU blue economy*. Publications Office of the EU. https://op.europa.eu/en/publication-detail/-/publication/79299d10-8a35-11e8-ac6a-01aa75ed71a1.

European Commission. (2019). *Common Maritime Agenda (CMA) for the Black Sea*. https://black-sea-maritime-agenda.ec.europa.eu/.

European Commission. (2021). *COM (EU) 2021/240 communication from the commission to the European parliament, the council, the European economic and social committee and the committee of the regions on a new approach for a sustainable blue economy in the EU transforming the EU's blue economy for a sustainable future*. https://eur-lex.europa.eu/legal-content/EN/TXT/?uri=COM:2021:240:FIN.

European Commission. (2022). *The EU blue economy report 2022*. Publications Office of the EU. https://op.europa.eu/en/publication-detail/-/publication/156eecbd-d7eb-11ec-a95f-01aa75ed71a1/language-en

European Commission. (2023). *Consequences of climate change*. https://climate.ec.europa.eu/climate-change/consequences-climate-change_en.

FAO. (2023). *Recent boom in aquaculture under threat in the Black Sea region | General fisheries commission for the Mediterranean (GFCM)*. Food and Agriculture Organization of the United Nations. https://www.fao.org/gfcm/news/detail/en/c/1506130/.

Geels, F. W., & Schot, J. (2007). Typology of sociotechnical transition pathways. *Research Policy*, *36*, 399–417. https://doi.org/10.1016/j.respol.2007.01.003

Hossain, M., Leminen, S., & Westerlund, M. (2019). A systematic review of living lab literature. *Journal of Cleaner Production*, *213*, 976–988. https://doi.org/10.1016/J.JCLEPRO.2018.12.257

Korpinen, S., Meski, L., Andersen, J. H., & Laamanen, M. (2012). Human pressures and their potential impact on the Baltic Sea ecosystem. *Ecological Indicators*, *15*(1), 105–114. https://doi.org/10.1016/J.ECOLIND.2011.09.023

Leminen, S., & Westerlund, M. (2017). Categorization of innovation tools in living labs. *Technology Innovation Management Review*, *7*(1), 15–25.

Loorbach, D. A., Rotmans J., & Wieczorek, A. J. (2007). Managing transitions for sustainable development.

Meadows, D. H. (2008). Thinking in systems: A primer. *Chelsea Green Publishing Company*, *133*(37), 14840. https://books.google.com/books/about/Thinking_in_Systems.html?hl=el&id=CpbLAgAAQBAJ.

Mendelow, A. L. (1981). Environmental scanning—The impact of the stakeholder concept. *Association for Information Systems AIS Electronic Library (AISeL)*. http://aisel.aisnet.org/icis1981/20.

Milkoreit, M. (2017). Imaginary politics: Climate change and making the future. *Elementa*, *5*. https://doi.org/10.1525/ELEMENTA.249/112449.

Ogonowski, C., Ley, B., Hess, J., Wan, L., & Wulf, V. (2013). *Designing for the living room: Long-term user involvement in a living lab*.

Riedy, C., & Waddock, S. (2022). *Imagining transformation: Change agent narratives of sustainable futures*. https://doi.org/10.1016/j.futures.2022.103010.

Rodrigues, M., & Franco, M. (2018). Importance of living labs in urban entrepreneurship: A Portuguese case study. *Journal of*

Cleaner Production, 180, 780–789. https://doi.org/10.1016/J.JCLEPRO.2018.01.150

Roorda, C., Wittmayer, J., Henneman, P., van Steenbergen, F., Frantzeskaki, N., & Loorbach, D. (2014). *Transition management in the urban context: Guidance manual*. https://researchbank.swinburne.edu.au/items/5d19a64e-5a34-46ef-832d-d1bdd8475f3d/1/.

The World Bank. (2020). *Bulgaria: Toward blue economy development*. https://documents1.worldbank.org/curated/en/750341608100452940/pdf/Toward-Blue-Economy-Development.pdf.

Trading Economics. (2023). *Georgia GDP from transport*. https://tradingeconomics.com/georgia/gdp-from-transport.

Voytenko, Y., McCormick, K., Evans, J., & Schliwa, G. (2016). Urban living labs for sustainability and low carbon cities in Europe: Towards a research agenda. *Journal of Cleaner Production, 123*, 45–54. https://doi.org/10.1016/J.JCLEPRO.2015.08.053

Yasuoka, M., Akasaka, F., Kimura, A., & Ihara, M. (2018). Living labs as a methodology for service design—An analysis based on cases and discussions from a systems approach viewpoint. In *DS 92: Proceedings of the DESIGN 2018 15th International Design Conference* (Vol. 1, pp. 127–136). https://doi.org/10.21278/IDC.2018.0350.

Multi-disciplinary Approach: From Soil Amendments to Food Systems

Assessment of Halophyte Biochar for the Amendment of Sandy Soil

Geetha Kannan, Monica isukapall, Meenakshi Sivaramakrishnan, B. G. Prakash Kumar and Ramachandran Subramanian

Abstract

Soil is one of the essential components required for plants to grow. Soil amendment becomes a necessity for poor and marginal lands to be brought into agricultural applications. The methods involving inorganic fertilizers are very common but not sustainable due to environmental impacts. A sustainable amendment in the form of biochar for soil is gaining popularity with time because of its carbon sequestration ability, high resident time in soil, and ability to enhance soil quality with minimal negative impact. In the current study, biochar made from Napier grass grown with saline water as a tool for soil amendment was explored for sandy soil, which has lower nutrient content and water holding capacity. Under optimized preparation, biochar was produced from the Napier grass grown with saline and sweet water. The effect of this biochar on the pH, electrical conductivity, water holding capacity, porosity, and cationic exchange capacity on the desert soil was studied to check its suitability for soil amendment and the results were compared with commercial potted soil available in the market. The results indicated a significant improvement in the physical properties of the sandy soil and were comparable to the commercial potted soil. Effect of the biochar on plant growth was studied in a pot experiment conducted for 90 days in which cow pea plant was grown in pots containing desert soil and soil amended with 10% and 20%(W/W) biochar. Analysis of variance (ANOVA) was used to identify significant differences in plant growth between the different mediums used for growth. Linear regression was also used to evaluate the predictive power of the variables identified by ANOVA.

Keywords

Biochar · Soil amendment · Sustainable · Water holding capacity · Cationic exchange capacity · Desert soil

1 Introduction

Population increase the world has seen has increased the demand on food production. Successful agriculture requires good soil which provides a conducive environment for plant growth. Soils of arid and semi-arid regions are usually low in organic matter and are thus deficient in nitrogen (Faradiella et al., 2015). Agriculture in a typical sandy soil found in arid and semi-arid areas requires frequent application of fertilizers and it is also highly water intensive in nature because of the soil's poor water holding capacity. (Abdullah & Shagufta, 2016). Soil amendment is hence a necessity to convert infertile lands to fertile ones for agricultural purposes. The industrial agriculture has increased our reliance on inorganic fertilizer as a soil amendment tool. The negative impacts of these fertilizers on the environment and human health clearly indicate that it is not a sustainable way forward. A sustainable amendment in the form of biochar for soil is gaining popularity with time (Rodriguez et al., 2009). Biochar, a carbonaceous product

G. Kannan (✉)
Department of General Sciences, BITS Pilani Dubai Campus, Dubai, United Arab Emirates
e-mail: geetha@dubai.bits-pilani.ac.in

M. isukapall · M. Sivaramakrishnan · B. G. Prakash Kumar
Department of Chemical Engineering, BITS Pilani Dubai Campus, Dubai, United Arab Emirates
e-mail: f20160334@dubai.bits-pilani.ac.in

M. Sivaramakrishnan
e-mail: f20150225d@alumni.bits-pilani.ac.in

B. G. Prakash Kumar
e-mail: prakash@dubai.bits-pilani.ac.in

R. Subramanian
Department of Biotechnology, BITS Pilani Dubai Campus, Dubai, United Arab Emirates
e-mail: sramachandran@dubai.bits-pilani.ac.in

M. Amer et al. (eds.), *Territorial Landscapes Solutions for Sustainable Energy and Food Security*,
Sustainable Landscape Planning and Natural Resources Management, https://doi.org/10.1007/978-3-031-82419-7_7

of biomass pyrolysis has the ability to increase soil pH (Lehmann et al., 2006), improve CEC (Chan et al., 2007), and improving water holding capacity (Bruno et al., 2002). Biochar is renewable and hence has a better carbon sequestration potential (Kharel et al., 2019). In tropical region, the effects of biochar on clayey soil (Lu et al., 2014; Obia et al., 2018) are well analyzed unlike Sandy soil. Data on the impact of biochar on soil characteristics and crop responses are very limited for sandy soil. Fresh water resources are fast depleting while brackish water and saline water availability are increasing due to water treatment process like desalination. In the current study, the biochar production from Napier grass grown with saline water was optimized and the resultant biochar was evaluated for the remediation of sandy soil for agricultural productivity.

2 Materials and Methods

2.1 Characteristics of Soils

Torripsamment dominates the soil map of UAE and hence for the study it was used. The topsoil (0–20 cm) was collected, air-dried, sieved (<2 mm) and stored for the studies. The soil sample was analyzed for its texture, pH, electrical conductivity, total organic matter, and cationic exchange capacity. It was a fine sand with a pHs of 7.5. Its electrical conductivity was 2.60 dS $m-^{1}$ while the total organic matter very low (about 0.2%). The cation exchange capacity was 3.3 $cmol_c$/kg.

2.2 Biochar Preparations

The Feedstock used for biochar production is Napier grass (Pennisetum purpureum), a fast-growing, deeply rooted perennial grass with excellent drought tolerance and soil stabilization characteristics. It is commercially used as fodders for dairy farming. Napier grass grown with fresh water (N1FW), 10% saline water (N1S1), and 20% saline water (N1S2) were harvested and dried. Dried grasses were cut, ground, and sieved to particles less than 2 mm in size.10 g of the dried biomass was taken in a crucible, and the pyrolysis temperature was varied from 150 to 500 °C under limited air flow for 180 min, and the resultant yields were noted. Maximum biochar yield was obtained when heated from 230 to 250 °C for 180 min at a heating rate of 1.3890 °C/min in a self-regenerated environment (Fig. 1). Biochar production was carried out with and without nitrogen passing through the reactor during pyrolysis. The presence and absence of nitrogen an insignificant difference in biochar yields. All the biochar used for the studies was thus prepared at the optimized condition.

2.3 Result and Discussion

The surface functionality of biochar was measured by Fourier Transform Infra-Red spectroscopy (FTIR). The chars were pelletized with KBr (1:200 (w/w)) and analyzed using a Spectrum BX II, Perkin Elmer with the software Spectrum 5.3 FTIR spectrophotometer (Fig. 2).

With the increase in temperature the –OH stretching from the water molecule reduced in intensity clearly indicating the dehydration process. Aliphatic –CH stretching is stronger in the biochar produced at 200 °C but get reduced in its intensity as the temperature increases but did not disappear indicating the presence of stronger aliphatic structure. The biochar produced a stronger C=O stretching which is observed around 1600 cm^{-1}. The maximum percentage yield of biochar was produced around 250 °C and at that temperature range, it was found to be more polar with functional groups such as –OH, C=C, C–O (Zambon et al., 2016). Biochar produced at low temperature has a less-condensed carbon structure and it helps to improve the

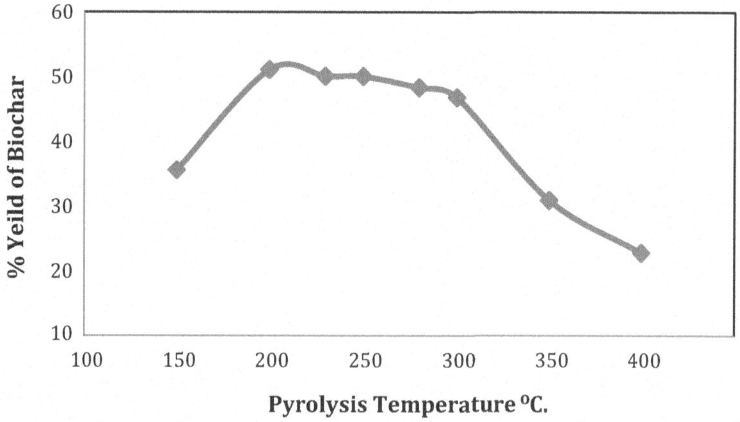

Fig. 1 Effect of pyrolysis temperature on the biochar yield

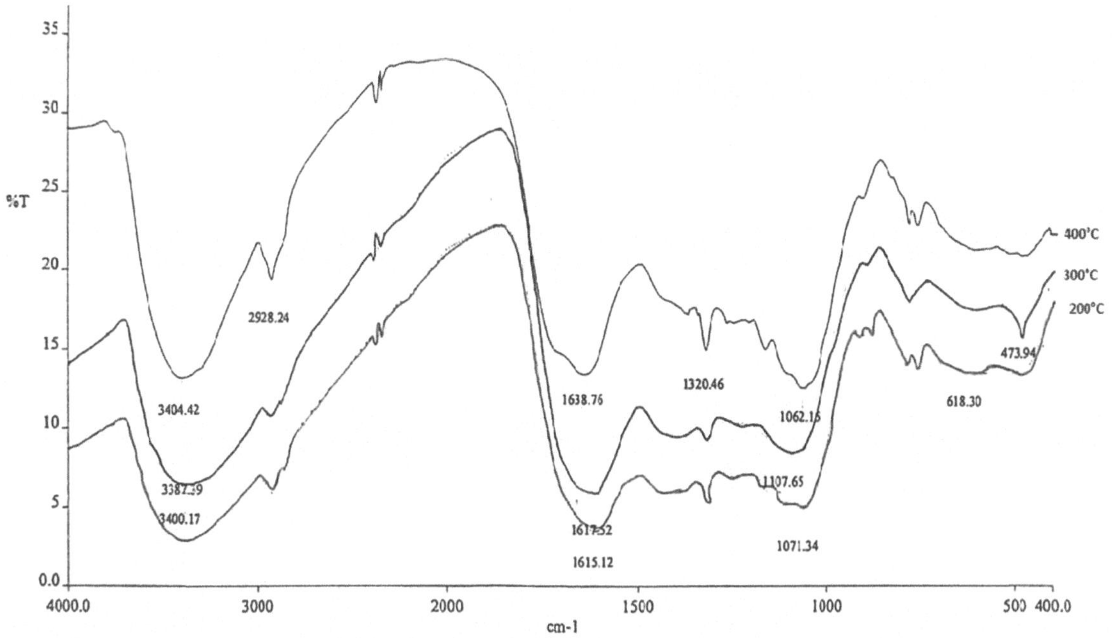

Fig. 2 FTIR Spectra of biochar obtained at different pyrolysis temperature

Table 1 Effect of biochar on the pH of desert and potted soil

S. no		pH	
		Desert soil (DS)	Potted soil (PS)
1	No biochar	7.5±0.01	7.38±0.02
2	With 10%N1S1	8.1±0.02	7.98±0.01
3	With 20%N1S1	8.37±0.01	8.28±0.01
4	With 10%N1S2	7.96±0.01	7.76±0.02
5	With 20%N1S2	8.2±0.02	7.93±0.01
6	With 10%N1FW	8.25±0.01	8.1±0.02
7	With 20%N1FW	8.4±0.01	8.3±0.02

fertility of soils. There was no significant difference among the three biochar produced, indicating clearly that the three biomass prepared were structurally similar.

The effect of biochar on soil was evaluated by mixing 0–20% (w/w) biochars (N1S1, N1S2, and N1FW) with desert soil (DS) and compared with 0–20% (w/w) biochars (N1S1, N1S2, and N1FW) with potted soil (PS). Water retention capacity, pH, Electrical conductivity, Cationic exchange capacity, and porosity were determined for the soil samples. pH of the desert soil was 7.5 and when amended with biochar, it increased. The difference in soil pH when amended with biochar produced from biomass raised with saline water and that of fresh water was insignificant. The pH values of the potted soil amended with biochars were comparable to that of desert soil amended with biochar (Table 1).

Electrical conductivity was expected to increase with soils amended with biochars made from biomass grown with saline water. It was found to be a reality but the magnitude increase was not very significant to bring any detrimental effect on the soil. The potted soil showed a higher conductivity than the desert soil (Table 2).

Desert soil amended with biochar showed a pronounced increase in the water retention capacity and was comparable to that of potted soil. Increased weight of biochar increased the water retention in desert soil whereas it was not that significant with potted soil (Table 3). The amendments reduced the porosity of the soil but there was no significant difference between the three biochar clearly indicating that the biochar from saline water grown biomass can play a positive role in increasing the water retention capacity which will prove beneficial on a longer run for the soil in semi-arid regions.

Table 2 Effect of biochar on the EC of desert and potted soil

S. no		EC (dS/m)	
		Desert soil (DS)	Potted soil (PS)
1	No biochar	0.00192	0.00307
2	With 10%N1S1	0.0042	0.0085
3	With 20%N1S1	0.0047	0.0088
4	With 10%N1S2	0.00398	0.00823
5	With 20%N1S2	0.0054	0.0086
6	With10%N1FW	0.00376	0.0047
7	With 20%N1FW	0.0036	0.0054

Table 3 Effect of biochar on the water retention and pore size of desert and potted soil

S. no		% Increase in water retention		% Reduction in pore size	
		Desert soil (DS)	Potted soil (PS)	Desert soil (DS)	Potted soil (PS)
1	With 10%N1S1	51	50	8	7
2	With 20%N1S1	89	75	12	10
3	With 10%N1S2	68	61	7	8
4	With 20%N1S2	92	79	11	10
5	With 10%N1FW	50	51	7	6
6	With 20%N1FW	90	81	12	11

Table 4 Effect of biochar on the CEC of desert

S. no		CEC (meq/100 g of soil)
		Desert soil (DS)
1	No biochar	3.1
2	With 10%N1S1	3.5
3	With 20%N1S1	3.64
4	With 10%N1S2	3.66
5	With 20%N1S2	3.78
6	With 10%N1FW	3.4
7	With 20%N1FW	3.56

Cationic Exchange Capacity (CEC) reflects on the soil's ability to retain the positively charged cations. CEC of the Desert soils was low and hence an attempt was made to check the CEC after amending with biochars. Biochar amendment did increase the CEC but only marginal increase was observed with an increase in biochar content (Table 4). CEC of the soil amended with N1S1, N1S2, and N1FW were comparable. The CEC was not measured for the potting soil, as the protocol for measuring the CEC for potted soil was different from that of soil and the unit of reporting also differ for the soil (meq per 100 g of soil) and potting soil (meq per liter of potting mix).

3 Plant Growth Study

3.1 Growing Substrates

A completely randomized design (CRD) with four replications was used in the experiment using black-eyed peas as the test crop. The treatment layout was a factorial combination of three different rates (0, 10, and 20 g kg−^{1}) of biochar added to the desert soil and 100% potted soil for reference. As there was no significant difference in the physical and chemical properties of different biochars, the plant growth study was conducted with biochar produced from 20% N1S2. The biochar rates were 0, 25, and 50, which corresponded to 0, 10, and 20 t ha^{-1}, respectively. The growth of the plants was analyzed for 90 days and all the parameters were kept constant except the biochar proportion in soil. The pots were kept outside in BITS Pilani Dubai campus building. The temperatures varied from about 40 °C in the morning to 30 °C in the night in the start of the experiment and by the end to 30 °C in the morning to 16 °C in the nights. Polyethene cover pieces were kept in the bottom of the pots to ensure there was no nutrient seepages during watering of the plants. The plants were watered 10 ml every day morning. All the plants had 4 replicates kept nearby to get more reliable results.

3.2 Growth Parameters

The number of leaves, diameter of the stems, height of plant, number of branches, and length of root were the parameters recorded. The changes in these parameters were measured every seven days. The plant height was measured from the tip of the shoot to the base of the plant. The first three leaves from the tip of the shoot were selected for the studies. The lower leaves were not considered as the probability of withering was more. The number of branches were counted from the primary stalk. All lengths were measured in centimeters.

3.3 Statistical Analysis

Completely randomized design (CRD) was used to analyze the results. The mean values of the different parameters were put in linearized models as the dependent variables and number of days as the independent variable. To determine if there is a significant change in plant growth with a change in biochar concentrations, 1-way ANOVA was used. The null hypothesis for ANOVA calculations was that there is no significant difference in the plant growth parameters among the soils with different concentrations of biochar.

The alternate hypothesis is that there is a significant difference in the plant growth parameters among the soils with different concentrations of biochar. The quantification of the changes observed was done using regression analysis. The different slope values show the strength of the relationship between the parameter and time. Greater the slope value, stronger the relationship between the growth parameter and the soil concentration. All tests were conducted with $P=0.05$ and were all done on MATLAB package with machine learning software 18.

4 Evaluation of Impact of Biochar on Plant Growth

4.1 Number of Leaves

The ANOVA results (Fig. 3) showed that the alternate hypothesis is clearly favoured as there is a positive change in the growth parameters with increase in biochar levels in soil. Plant growth in soil amended with 20% N1S2 biochar was almost equal to that observed with potted soil. The ANOVA results showed that the maximum, minimum, and average values for the plant grown in 20% biochar char were all equal or greater than the potted soil. The spread of the results is also supportive of the hypothesis, the 20% biochar has a greater spread of values which shows that the range of values for the number of leaves is larger than potted soil. The consistency of results is not as strong as in potted soil but the difference in results was not significant therefore it could be said that they are both at par with each other. Considering the 10% biochar plant the spread is less therefore consistent results in all pots but the maximum and minimum values were less efficient than with 20% biochar. Plot of number leaves vs time showed that the slope in 20% biochar (Fig. 4) was almost equal therefore showing that both 20% and potted soil have the same strength between the parameter and time.

4.2 Diameter of Stems

Diameter of stems for the 20% biochar pots showed a large range of values indicating low consistency on the other hand, potted soil has a smaller range of values indicating more consistence. Nevertheless, the average values of potted soil and 20% biochar around the same value making up for the irregularity (Fig. 5). Results with 10% biochar pots have a lower average value but was more consistent than 20% biochar. The regression slope (Fig. 6) values are almost the same for 20% biochar and potted soil Supporting the hypothesis that desert soil amended with 20% biochar gives a larger diameter of stems almost as equal to fertilized potted soil.

4.3 Height of the Plant and the Number of Branches

A clear distinction in the different values of the heights was seen when the biochar concentrations were varied in the plants. As expected, the 20% biochar plants and the potted soil plants showed the highest average values. It was also

Fig. 3 ANOVA result of number of leaves

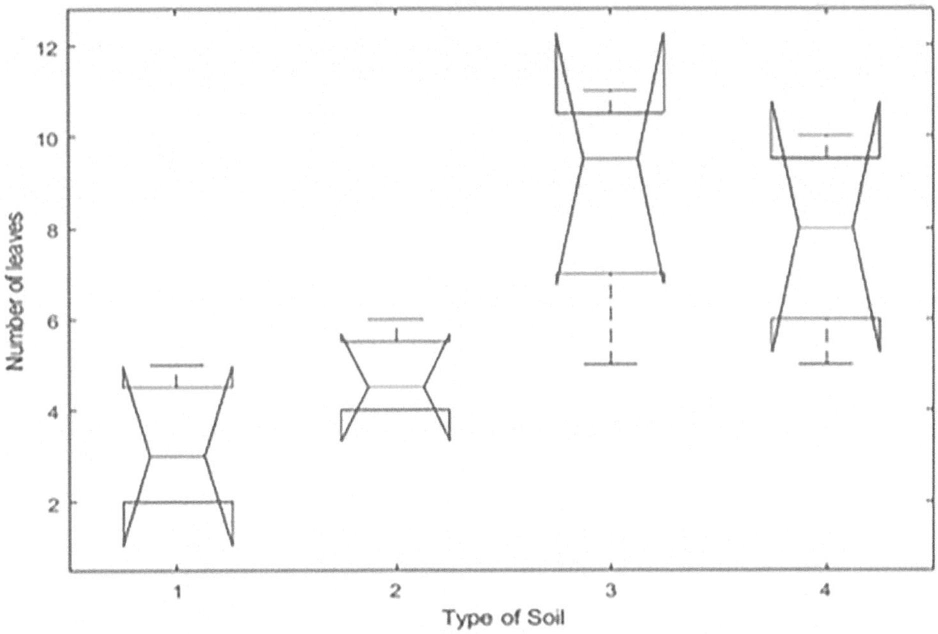

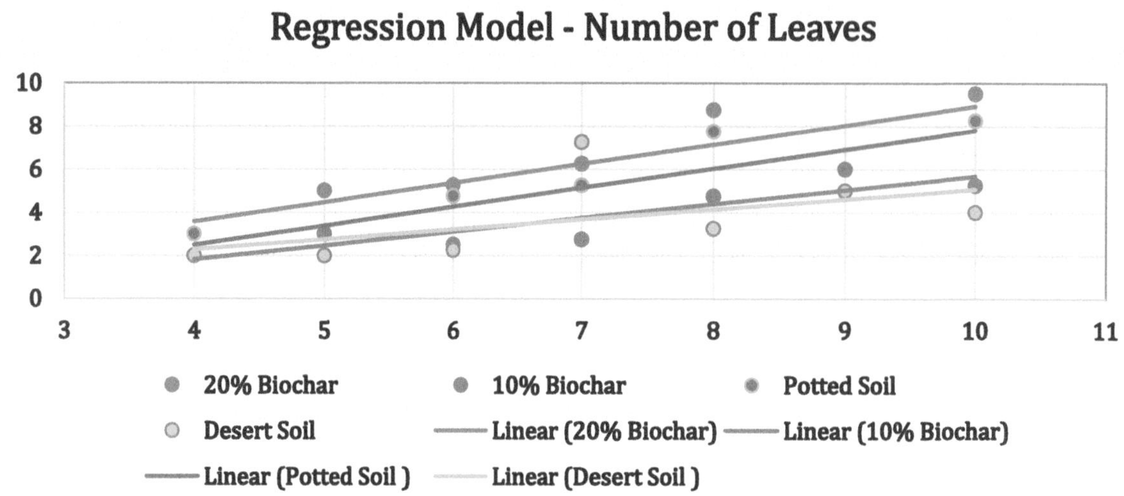

Fig. 4 Regression results for number of leaves

Fig. 5 Diameter of stems ANOVA results

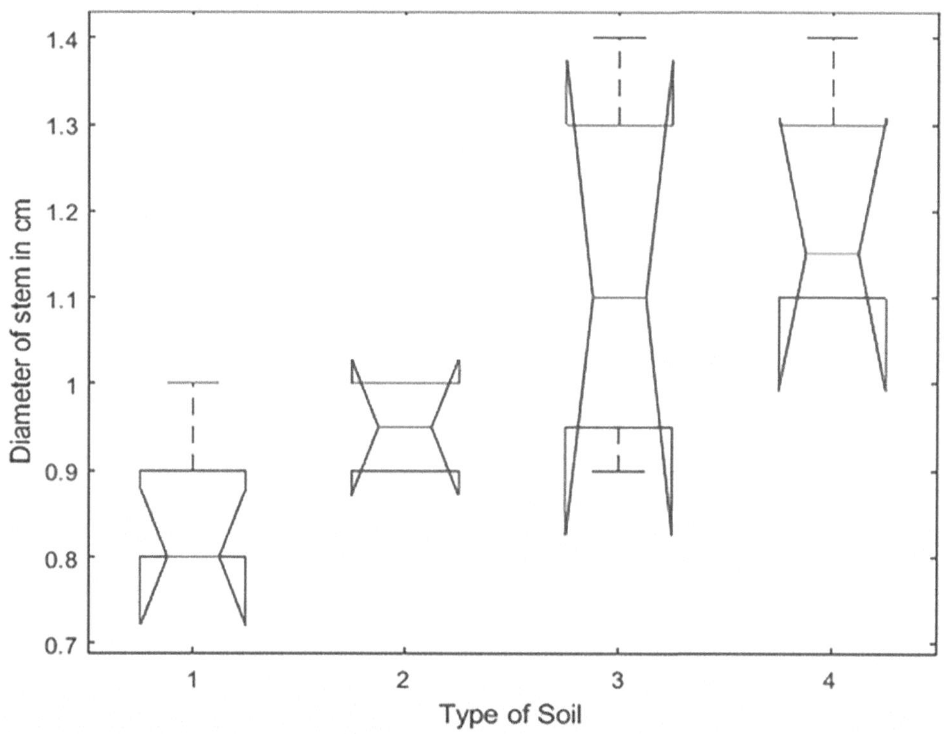

observed that the spread of results in both 20% and potted soil was very less only ranging between 17< x <20cm (Fig. 7). This showed that the biochar addition gave consistent growth in terms of the height of the plants like fertilized potted soil. While on the other hand, Desert soil plants had very vast variation of values. This was improved by the addition of biochar and plant growth was also increased. To emphasize further, the height of the plants with 20% biochar in the soil showed a greater rate than potted soil plants as seen from the regression slope values (Fig. 8). Similar results were observed for the parameters like number of branches (Figs. 9 and 10) and the size of the leaves (Fig. 11). To conclude, for all the growth parameters, not only does the growth rate increased there was also consistent with biochar amendment.

4.4 Length of Roots

It was observed that the biochar applied soil had longer roots than the fertilized potted soil or regular desert soil. This is unconventional because we would expect the fertilized soil

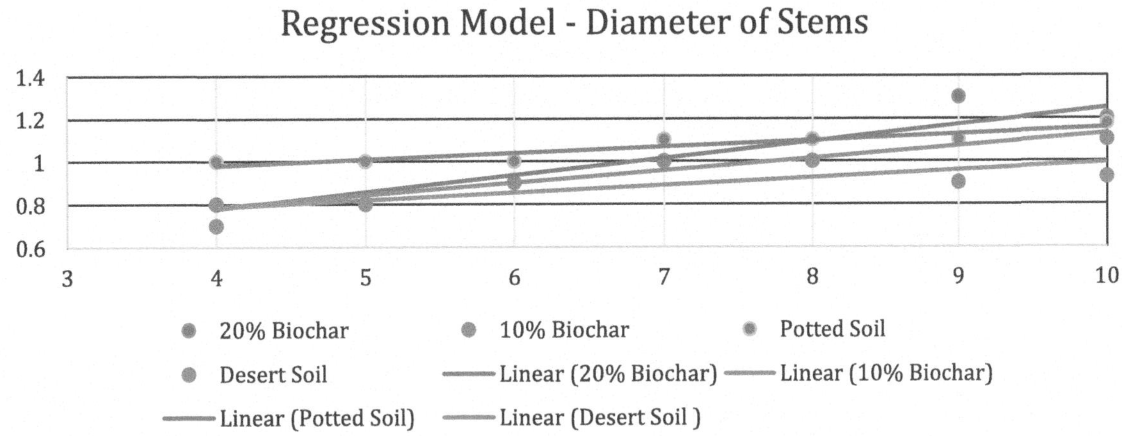

Fig. 6 Regression results of diameter of stems

Fig. 7 ANOVA results of height of the plants

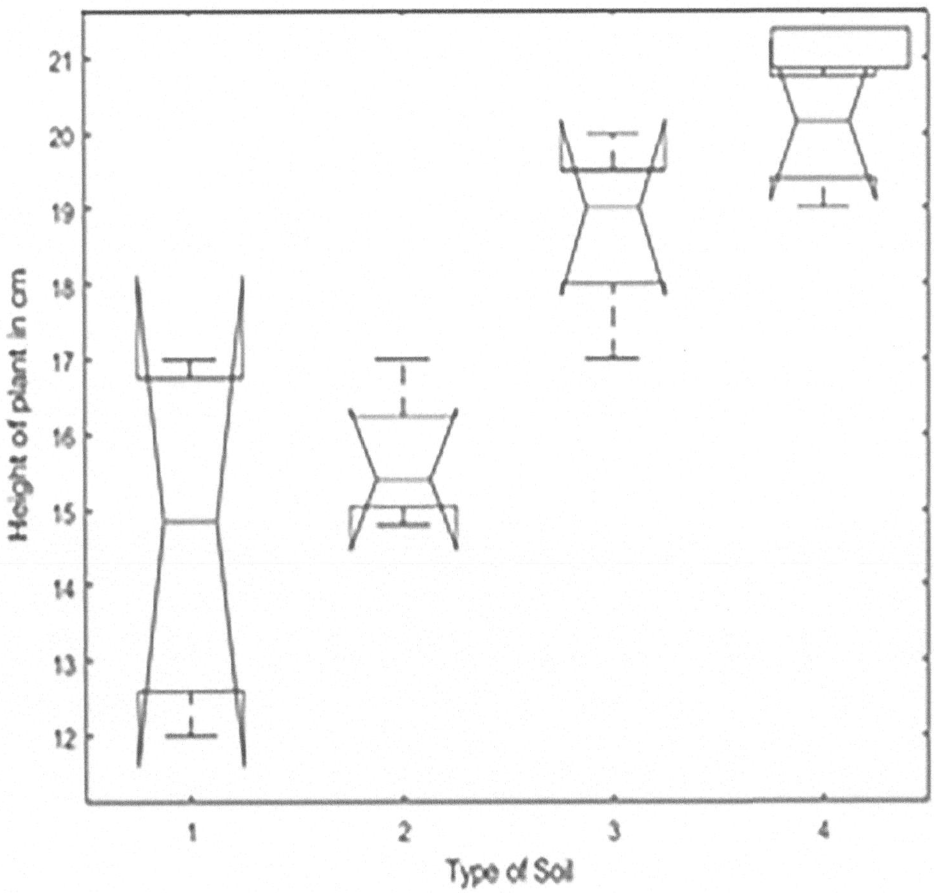

to have longer stronger roots as more nutrients are available. The results show quite the opposite, 10% biochar had the longest roots and they are very spread out. There was a main Tap present and the vast spread of roots around it. The 20% biochar roots are slightly smaller than the 10% biochar roots, this could be because the nutrients are more easily available for these plants than 10% as there is more facilitator. More the nutrients then smaller the root length is a conclusion that can be made from the root length and the different biochar compositions. Longer and larger area of roots make the plants more grounded and stable than smaller roots therefore adding biochar also increases the stability of the plants.

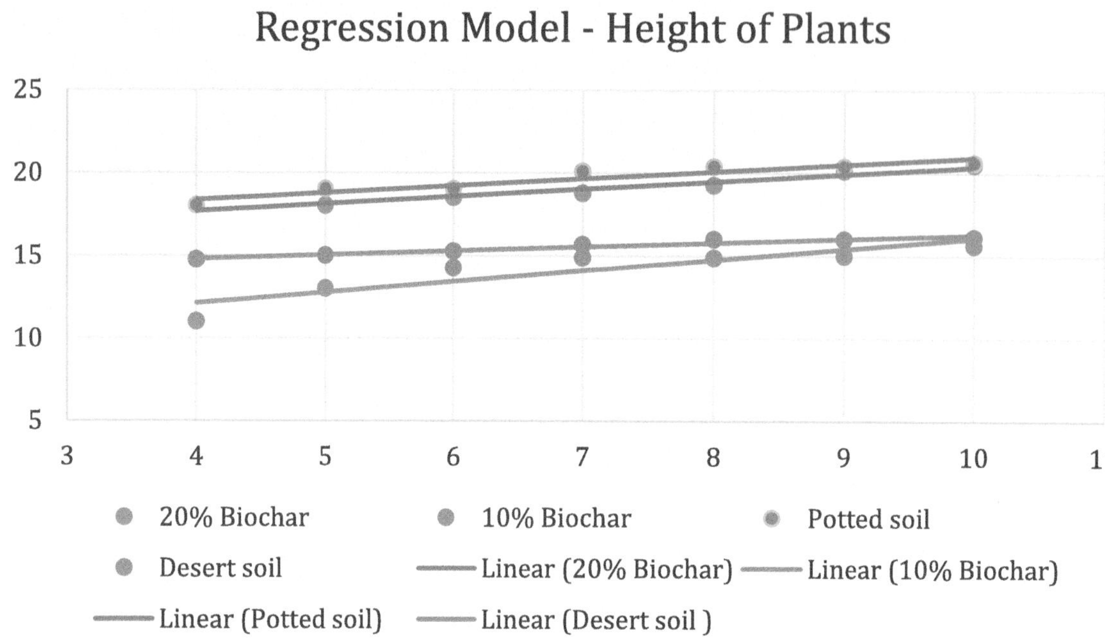

Fig. 8 Regression results of height of the plants

Fig. 9 ANOVA results for number of branches

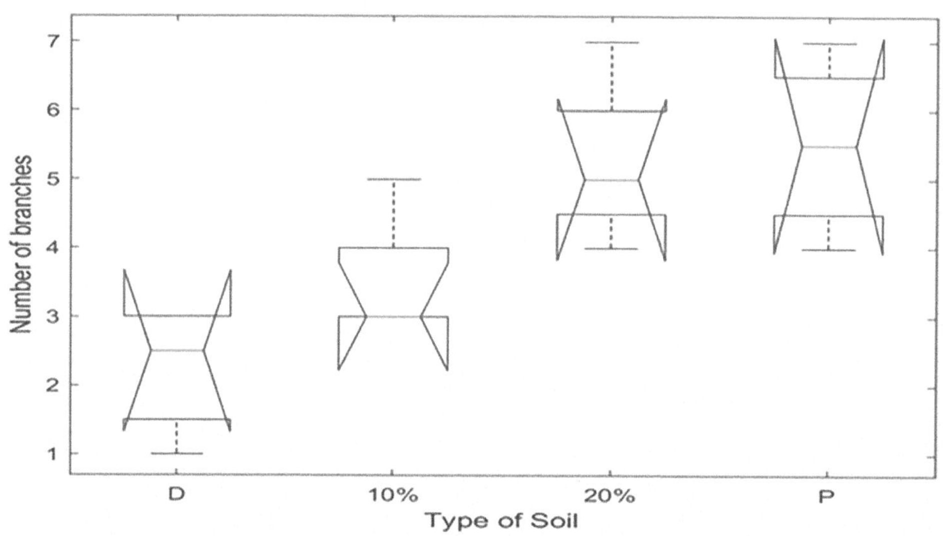

5 Conclusion

Agriculture makes a heavy demand on the already depleting fresh water resources around the world. If saline water could be used it would reduce the demand on the fresh water supply. Napier grass grown with saline water was effectively used to produce biochar which when amended to the desert soil, it improved the pH, electrical conductivity, cationic exchange property, and water retention property. The study explicitly showed that desert soil amended with biochar produced from saline water grown biomass was equivalent to that grown with fresh water. The pot study on plant growth clearly indicated that the desert soil amended with biochar was having all the beneficial characteristics as that of commercial potted soil.

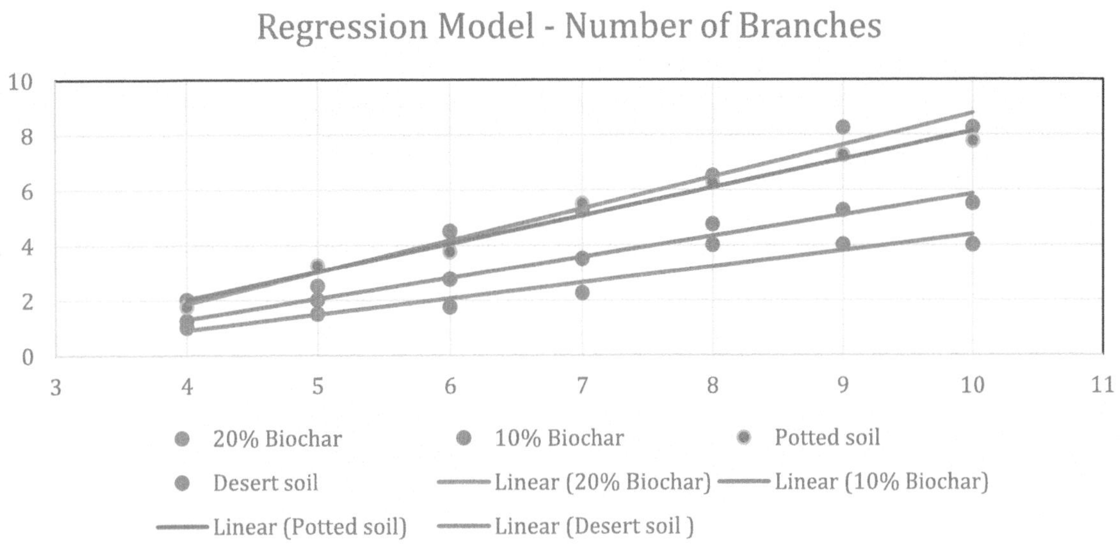

Fig. 10 Regression results of number of branches

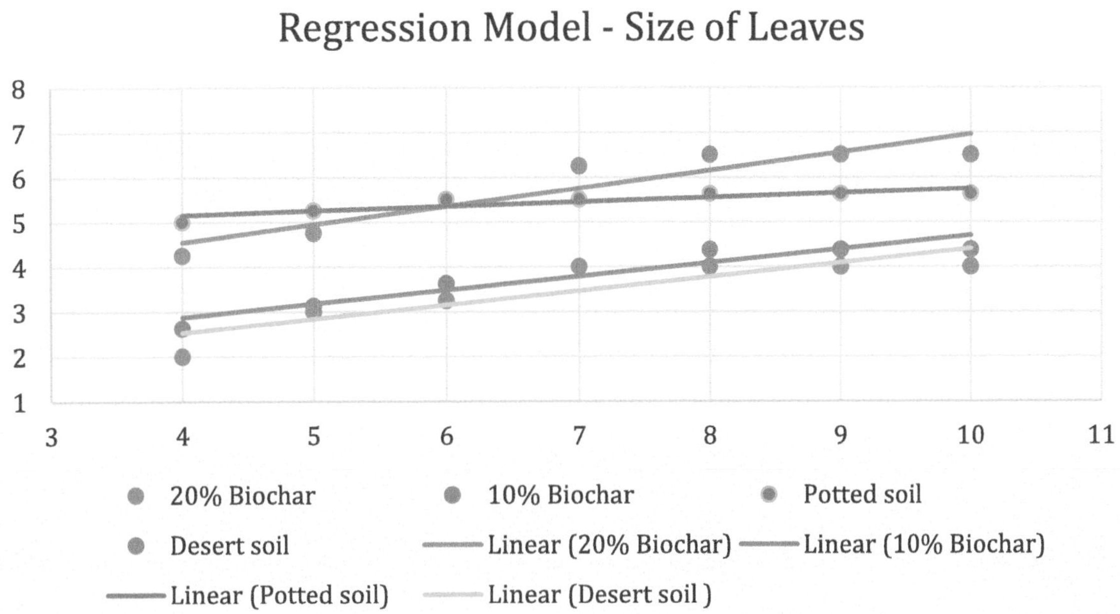

Fig. 11 Regression results on the size of leaves

References

Abdullah, A. S., & Shagufta, G. (2016). Integrated plant nutrient management for sandy soil using chemical fertilizers, compost, biochar and biofertilizers case study in UAE. *Journal of Arid Land Studies, 26–3*, 101–106.

Bruno, G., Lehmann, J., & Zech, W. (2002). Ameliorating physical and chemical properties of highly weathered soils in the tropics with charcoal—A review. *Biology and Fertility of Soils, 35*, 219–230.

Chan, K.Y., Van Zwieten, L., Meszaros, I., Downie, A., & Joseph, S. (2007). Agronomic values of green waste biochar as a soil amendment. *Soil Research, 45*, 629–634.

Mohd Kusin, F., Izzati, N., Mohamat-Yusuff, F., Awang, M. (2015). The impact of nitrogen fertilizer use on greenhouse gas emissions in an oil palm plantation associated with land use change. *Atmósfera, 28*(4), 243–250.

Kharel, G., Sacko, O., Feng, X., Morris, J. R., Phillips, C. L., Trippe, K., Kumar, S., & Lee, J. W. (2019). Biochar surface oxygenation by ozonization for super high cation exchange capacity. *Sustainable Chemistry and Engineering, 7*(19), 16410–16418.

Lehmann, J., Gaunt, J., & Rondon, M. (2006). Biochar sequestration in terrestrial ecosystems—A review. *Mitigation Adaptation Strategies for Global Change, 11*, 403–427.

Lu, S. G., Sun, F. F., & Zong, Y. T. (2014). Effect of rice husk biochar and coal fly ash on some physical properties of expansive clayey soil (Vertisol). *CATENA, 114*, 37–44.

Obia, A., Mulder, J., Hale, S. E., Nurida, N. L., & Cornelissen, G. (2018). The potential of biochar in improving drainage, aeration and maize yields in heavy clay soils. *PLoS ONE, 13*(5), e0196794.

Rodriguez, L., Salazar, P., & Preston, T. R. (2009). Effects of biochar and digester effluent on growth of maize in acid soils. *Livestock Research for Rural Development, 21*(7), 110.

Zambon, I., Colosimo, F., Monarca, D., Cecchini, M., Gallucci, F., Proto, A. P., Lord, R., & Colantoni, A. (2016). An innovative agro-forestry supply chain for residual biomass: Physicochemical characterisation of biochar from olive and hazelnut pellets. *Energies, 9*(7), 526.

Scope of a Millet-based Food System in Reducing Metabolic Syndrome and Carbon Footprint in the UAE: A Policy Perspective

Subhasree Ray and Shoba Suri

Abstract

The United Arab Emirates (UAE) faces a dual challenge of addressing rising metabolic syndrome (MetS) and reducing greenhouse gas emissions. Metabolic syndrome, marked by obesity, high blood pressure, and abnormal cholesterol levels, affects around 40% of the UAE's population. Sedentary lifestyles and increased Western diet consumption contribute to its prevalence. The country aims to enhance food security and combat lifestyle disorders through innovative investments in health and agriculture. The UAE is teaming up with the World Health Organization (WHO) to combat childhood obesity, recognizing the impact of early habits on adult health. Millets are a popular health-conscious choice due to their low glycemic indices and high fiber content, which can potentially prevent diabetes, hyperlipidemia, and obesity. Millet's abundant micronutrient profile makes it an ideal food to combat nutritional deficiencies. These smart crops provide environmental benefits as well. Millets require less water for cultivation, making them ideal for arid UAE conditions. They promote sustainable agriculture by diversifying crops, improving soil health, and producing fewer greenhouse gas emissions, which also benefits the environment. This policy perspective proposes a holistic approach to address Metabolic Syndrome and its environmental impact through incentivizing local millet farming, nutrition education campaigns, and research collaboration using millets to ensure sustainable food security and nutrition.

S. Ray (✉)
Nutrition & Wellness, Reliance Industries Limited, Mumbai, India
e-mail: subhasree.ray@ril.com

S. Suri
Observer Research Foundation, Health Initiatives, New Delhi, India
e-mail: shoba.suri@orfonline.org

Keywords

Millets · Sustainable food systems · Metabolic syndrome · Climate change · Smart agriculture

1 Introduction

The United Arab Emirates (UAE) is dealing with a two-fold challenge of tackling the increasing occurrence of metabolic syndromes (MetS) among its population and reducing the emission of greenhouse gases (GHGs). The UAE has made numerous innovative investments in technology, climate-smart agriculture, and optimizing food supply chains to boost food sovereignty and security. Additionally, in partnership with the World Health Organization (WHO), the UAE has created a national plan to fight childhood obesity since childhood is the most crucial phase that shapes dietary and lifestyle habits in adulthood. This policy perspective offers a holistic approach to how the UAE can utilize climate-smart crop millets to decrease the environmental impact and metabolic syndrome, while also working towards attaining sustainable food security and nutrition adequacy for the population.

2 The Dual Burden on Health and Environment—Current Status of Knowledge

Metabolic syndrome refers to a group of health conditions including obesity, high blood pressure, high blood sugar, and abnormal cholesterol levels. These conditions increase the risk of developing cardiovascular disease and diabetes. A 2021 study (Mahmoud & Sulaiman, 2021) showed that nearly 40% of the UAE population was affected by either prediabetes or diabetes. Furthermore, the rates of overall dyslipidemia (73%) and hypertension (31%) in the UAE were found to be higher than the global rates. Overall,

M. Amer et al. (eds.), *Territorial Landscapes Solutions for Sustainable Energy and Food Security*,
Sustainable Landscape Planning and Natural Resources Management, https://doi.org/10.1007/978-3-031-82419-7_8

a large sample study by the World Health Organization (Noncommunicable diseases, 2023) from all UAE residents reported a high prevalence of MetS (40%). In 2022, Mahmoud and Sulaiman reported (2021) a significant prevalence of MetS in the UAE, especially among Emirati women and Asian non-Arab men. The young adult population in the UAE has a higher occurrence of MetS compared to global estimates in the same age group. In recent years, there has been a rise in sedentary lifestyles with high energy consumption. The UAE Global School Health Survey (Haroun et al., 2018) reported a significant increase in the percentage of overweight and obese Emirati adolescents from 21.5% in 2005 to 39.2% in 2010. The country's economic boom followed by an influx of Western culture (Belal, 2009) led to a dietary shift towards refined carbohydrates, trans fat consumption, and low fiber intake is one of the major factors contributing to the growing prevalence of MetS. According to the Global Nutrition Report (GNR) (2024), The United Arab Emirates has made limited progress in achieving diet-related non-communicable disease (NCD) targets. The report highlights that 44.2% of adult women (18+years) and 30.9% of adult men in the UAE are suffering from obesity. This is higher than the regional average of 10.3% for women and 7.5% for men. Furthermore, around 17.4% of adult women and 17.3% of adult men in the UAE are affected by diabetes. The report also suggests that people in UAE consume fewer health-promoting food groups (fruits, vegetables, legumes, nuts, and whole grains) and more of the food groups that have negative effects on health and the environment (red meat, dairy, and fish) than recommended by EAT-Lancet Commission. Figure 1 illustrates these figures.

In 2021, the UAE generated 193.5 million metric tons of carbon dioxide ($MtCO_2$) emissions due to fossil fuel combustion and industrial purposes. This represents a two per cent increase from the previous year (Statista, 2023). The UAE has one of the highest CO_2 emissions per person globally, with over 20 metric tons emitted per capita (Fig. 2) (Statista, 2023). WHO 'Health & Climate Change Country Profile' 2019 stated (Health & climate change: country profile, 2019) that the UAE may experience extreme heat, storm surges, rising sea levels, water stress, dust and sandstorms, and desertification due to climate change. Even minor changes in weather patterns could have a major impact on the country's economy, environment, and social stability. The most susceptible areas in the UAE are its water, coastal, marine, and dryland ecosystems, buildings and infrastructure, agriculture and food security, and public health.

The United Arab Emirates is dealing with specific difficulties that make it more susceptible to the effects of climate change on food and nutrition security. These challenges include a growing population, fragile natural surroundings with limited arable land, a high risk of climate change, and a heavy reliance on imported food. Additionally, climate change is expected to worsen metabolic and lifestyle risk factors associated with non-communicable diseases related to diet.

Only 5% of the land in the UAE can be used to support agriculture. Hence, over 80% of the food consumed in the UAE is imported from various regions, resulting in a significant impact on the environment (https://www.emiratesnaturewwf.ae/en/food-water-security). With these changing climate scenarios, trade flows to the UAE may be restricted, particularly for cereals, vegetables, and fruits. This could lead to higher prices and a larger portion of household income being allocated to food. To adjust to climate change, the UAE is already implementing strategies (Emirates Nature-WWF, 2024) to broaden its food import sources and enhance its domestic agricultural industry and millets could be an excellent choice for building a climate-resilient food system for the growing population.

3 Millets—Good for Health, Good for the Environment, and SDGs

Small-seeded grasses known as millets are becoming increasingly recognized for their exceptional nutritional and health benefits. These ancient grains, which include

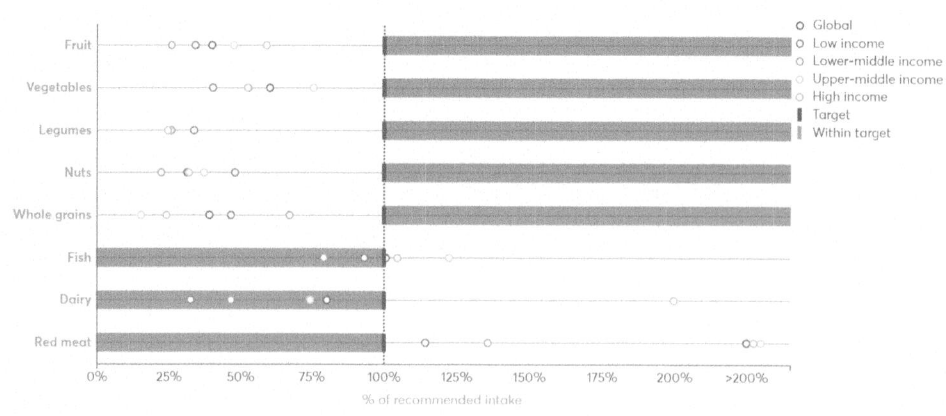

Fig. 1 Comparison of dietary intakes of key foods and nutrients among adults aged 20 and over with minimum and maximum targets—The UAE (GNR, 2024). *Source* Global Nutrition Report (GNR) (2024)

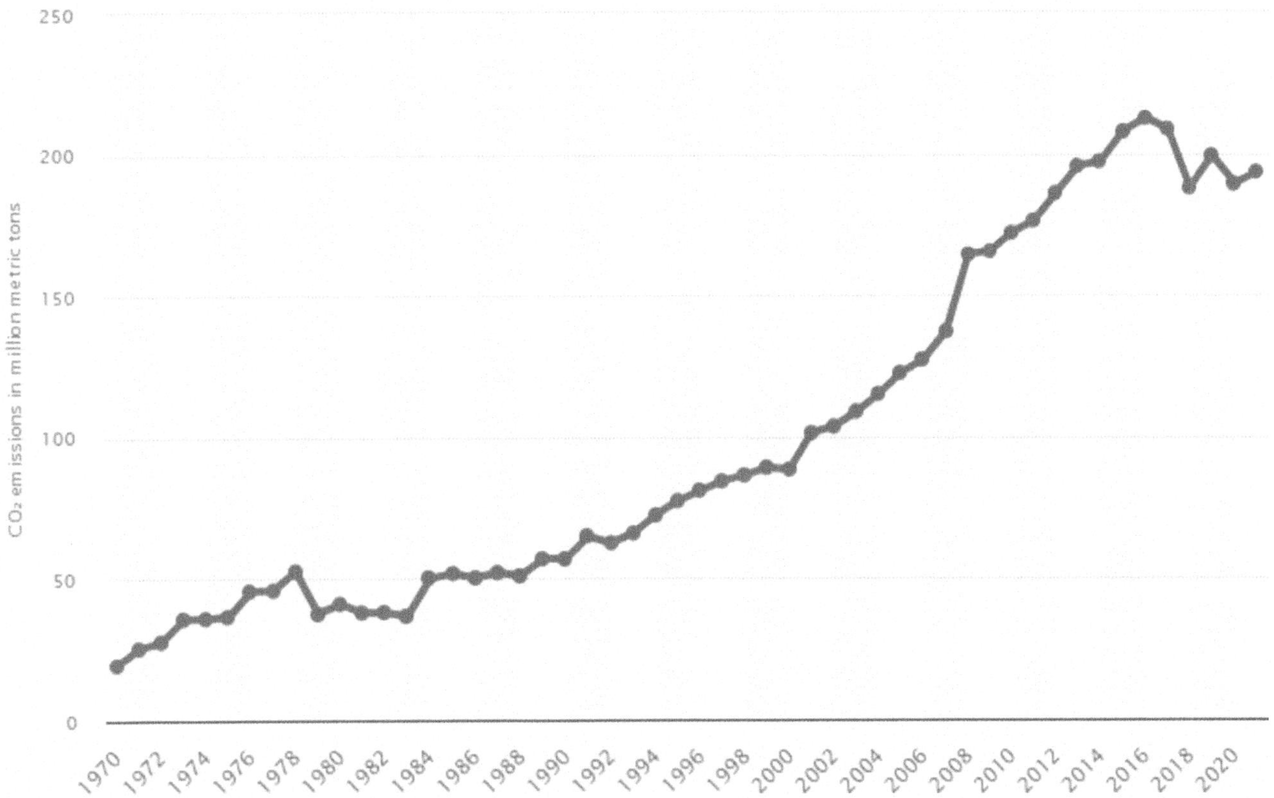

Fig. 2 Carbon dioxide emissions from fossil fuel and industrial purposes in the United Arab Emirates from 1970 to 2021 (in million metric tons). *Source* Statista (2023)

pearl millet, finger millet, sorghum, and foxtail millet, offer a wide range of advantages that contribute to overall well-being.

Millets are renowned for their impressive nutrient profiles, making them a rich source of essential nutrients such as dietary fiber, protein, vitamins (including niacin, B vitamins, and folate), and minerals (such as iron, magnesium, and phosphorus). These nutrients play a vital role in enhancing energy levels, improving digestion, and promoting better metabolic function.

Millets are highly regarded for their ability to manage and prevent chronic diseases (Climate et al., 2023). They have a low glycemic index and can regulate blood sugar levels, which has been linked to a decreased risk of type 2 diabetes. Studies have shown that including millets in one's diet can improve glycemic control and insulin sensitivity, two crucial factors in managing diabetes. A cross-sectional study (Rao et al., 2019) included 142 participants, 69 of whom followed a millet-based diet, and 73 of whom followed a non-millet-based diet. The results revealed that the millet-based diet group had a significantly lower BMI, systolic BP, FBS, PPBS, total cholesterol, LDL, and triglyceride levels, and higher HDL levels compared to the non-millet-based diet group. A systematic review (Anitha et al., 2021) of 19 studies was conducted to analyze the effects of consuming millets on lipid profile, including levels of total cholesterol (TC), triacylglycerol, high-density lipoprotein cholesterol (HDL-C), low-density lipoprotein cholesterol (LDL-C), and very–low–density lipoprotein cholesterol (VLDL-C). The review found that consuming millets for periods as short as 21 days to 4 months reduced levels of TC, triacylglycerol, LDL-C, and VLDL-C ($p<0.01$) by 8.0%, 9.5%, 10%, and 9.0%, respectively. Additionally, four studies showed that consuming millets brought TC and triacylglycerol levels to normal levels (<200 and <150 mg/dl, respectively). Moreover, millet-based meals resulted in a 6.0% increase in HDL-C, a 4.0 and 5.0% reduction in systolic and diastolic blood pressure, and a 7.0% reduction in body mass index (BMI). This evidence concludes that consumption of millets can be beneficial for preventing and managing metabolic syndrome.

Growing millet provides numerous environmental benefits (Anitha et al., 2022), making it a sustainable option for agriculture and diets. Millets are considered eco-friendly due to their low input needs and ability to grow in various climates. These robust crops are very tolerant of heat (up to 64 °C), drought-resistant, and require less water than water-intensive staples like rice and wheat. Millets are versatile crops that can thrive in a variety of environments, from low to high altitudes and across a wide range of latitudes. They can grow on arid lands without irrigation and can tolerate very low levels of rainfall. Additionally, millets have a low water footprint. This makes them suitable for areas facing water scarcity and climate change challenges.

Millet cultivation can promote sustainable agriculture by reducing dependence on synthetic fertilizers and pesticides. This can lead to crop rotation diversification and the avoidance of mono-cropping systems. Planting millets is beneficial for biodiversity and soil health (Millets—Good for people, the environment, and farmers, 2023). Their growth cycle and root structure help improve soil structure, lessen erosion, and boost soil organic matter. Additionally, millets have a smaller carbon footprint and generate fewer greenhouse gas emissions.

The United Nations General Assembly declared 2023 the International Year of Millet, responding to a proposal from India. With millet production in 21 states, India is the world's largest millet producer. A production volume of 173 lakh tonnes makes up 80% of Asia's millet production and 20% of global production. During India's presidency, a major focus of the G20 has been the 3S strategy for food security, which stresses 'smart' and 'sustainable' agriculture that should 'serve' all (Gururani et al., 2021). Many countries that are members of the G20, such as Saudi Arabia, Belgium, Germany, and Italy, are contributing the largest portions of millet imports. These nations have started to establish sustainable agriculture systems by introducing millet into their agricultural and dietary practices. Millet farming is also less expensive and requires less water usage, making it an attractive option for these countries.

Sustainable development goals (SDGs) can be met by promoting millet cultivation. One of the goals (Goal 2) is to achieve zero hunger. The UN General Assembly, 2015, set this goal as the goal of 'ending hunger, improving food security, improving nutrition, and promoting sustainable agriculture'. In today's agriculture, fertilizers and pesticides are almost never not used. The exception is millet. For thousands of years, millets have been grown as rain-fed crops with little or no fertilizer input (ORF, 2023). Also, millets do not require any pesticides (Devkota et al., 2016; Goron & Raizada, 2015; Gupta et al., 2017) because they are less vulnerable to insect attacks. One of the risk avoidance strategies in drought-prone areas of Africa is to store millets like finger millet seeds for several years without insect damage (Goron & Raizada, 2015; Saxena et al., 2018).

The most suitable crop for changing climates is pearl millet, which is highly tolerant to elevated temperatures and droughts (Ceasar, 2021). With their climate resilience properties, millets continue to promote sustainable agriculture without requiring chemical fertilizers (Saha et al., 2016). In the absence of indiscriminate use of fertilizers and pesticides in agriculture production, millet cultivation, and consumption can promote sustainable agriculture without harming the environment. Millets provide climate resilience features that contribute to the achievement of another Sustainable Development Goal (Goal 13; Climate Action). A major objective of SDG 13 is to increase the capacity of least developed countries and small island developing states to plan for and manage climate change. SDG 13 may be met in two ways by millets. The adverse effects of global warming and climate change could be overcome by millets, in one way. Furthermore, millet-based agriculture can reduce the production of synthetic fertilizers and pesticides, thus reducing environmental harm. Hence, millet can be a key factor in reducing fertilizer demand in the future and meeting food demand. Millet cultivation and value-added processing associated with millet production can also contribute to rural income generation, thus meeting SDG 1.

A steadily increasing consumer preference for millet due to lifestyle related illnesses also calls for an increase in millet production in the future (Thilakarathna & Raizada, 2015). SDG 3 (Good health and well-being) could also be achieved through millet consumption. The modern food system may lack essential nutrients. Changing consumer eating habits is crucial for increasing millet consumption. Many SDGs can be achieved through transformative changes in the food system, including the promotion of local foods like millet. Millet cultivation and consumption can contribute to achieving many UN SDGs since they are interconnected.

4 Millet-Centric Policy Perspective for the UAE

UAE is leading in innovative approaches to reduce environmental footprint and enhance the health profile of its citizens. The International Center for Biosaline Agriculture (ICBA) is currently conducting research on nutritious and resilient crops that can be used for dietary diversification in harsh environments like the UAE (Kane-Potaka et al., 2021). The project, titled "Evaluation of diverse and nutrient-rich, climate-resilient crops/accessions for dietary diversification in marginal environments", began in 2021 and involved evaluating ten different food crops under varying salinity conditions. These crops were chosen for their nutritional profile, resilience to environmental stressors like salinity, drought, and heat, and adaptability to different cropping systems. Preliminary results have identified four promising crops—finger millet, foxtail millet, pearl millet, and proso millet—that can contribute to a healthy diet. During the 2022 growing season, the researchers of ICBA conducted replicated trials on 20 cultivars of finger millet, foxtail millet, and pearl millet, as well as 25 cultivars of proso millet, under varying water salinity and irrigation conditions. The findings of this study are significant, particularly given the upcoming International Year of Millets in 2023. This aims to promote millet as a key component of the food basket and advance the implementation of SDGs 2, 3, 12, and 13. Ultimately, the study aims to provide data on the consistency and stability of these cultivars' performance on a large scale, with the best-performing cultivars from each crop to be encouraged for promotion and commercialization to boost food and nutrition security.

The current policy perspective recommends a multi-pronged approach to promote millet consumption to address both metabolic syndrome and environmental impact.

a. **Sustainable Agriculture**—Provide subsidies and incentives for farmers to grow millet crops, including financial support, training, and access to seeds. Develop and implement sustainable farming practices for millet production to minimize carbon emissions and water usage. Encourage crop rotation and diversification to improve soil health and reduce the need for chemical inputs. This will enhance food security, contribute to climate resilience, and reduce the environmental impact of importing water-intensive staples. Additionally, establishing community gardens and urban farming initiatives can engage citizens in sustainable millet cultivation, improving food security, promoting healthy lifestyles, and enhancing community resilience. Already established protocols such as vertical farming and organic farming can be leveraged to promote millet cultivation. Successful sustainable millet agriculture will require collaboration between governmental agencies, research institutions, and local communities.
b. **Food Production and Processing**—Promote the development of millet-based food processing industries to create a variety of millet-based products. Establish quality control and safety standards for millet-based products to ensure consumer trust. Encourage food manufacturers to label products as "millet-based" for easy identification. Furthermore, forging partnerships with food processing industries can create a market for millet-based products, strengthening the value chain and contributing to economic viability. Encourage the use of renewable energy sources in millet processing and transportation.
c. **Environmental Sustainability**—Promote organic farming practices to reduce the environmental impact of millet production. Encourage efficient water management and sustainable land use practices. Monitor and regulate the use of fertilizers and pesticides to minimize their impact on the environment.
d. **Nutrition Awareness**—Create engaging and informative content that showcases millets' nutritional benefits, versatility, and culinary uses. Collaborate with healthcare professionals such as doctors, nutritionists, and dietitians to educate the public on the health benefits of incorporating millets into their diets. Host workshops, webinars, and seminars to discuss metabolic health and the role of millet. Organize cooking demonstrations, food festivals, and workshops featuring millet-based dishes. Partner with renowned chefs and culinary schools to showcase millets' delicious and diverse preparations. Integrate millet awareness into school curricula to educate students on the nutritional value and environmental benefits of millet. Establish school gardens or green spaces to demonstrate millet cultivation and involve students in hands-on learning.
e. **Consumer Awareness and Demand**—Run public awareness campaigns to educate consumers about the benefits of millet-based diets and how they can contribute to reducing metabolic syndrome. Promote millet-based recipes and cooking workshops to encourage consumers to incorporate millets into their daily meals. Create dedicated websites, blogs, and social media pages to share information, recipes, success stories, and testimonials related to millets. Engage influencers and social media personalities to promote millets. Collaborate with local newspapers, magazines, and online publications to feature articles, interviews, and stories about millets, their benefits, and their potential to address metabolic health and sustainability.
f. **Government Procurement**—Include millet-based products in government-run institutions, schools, hospitals, and military canteens. Offer tax incentives and subsidies to institutions and businesses that adopt millet-based menus. Conduct regular assessments of the carbon footprint reduction achieved through millet-based procurement. Seek support from governmental agencies and policymakers to include millet awareness in health and nutrition programs, national campaigns, and sustainability agendas. Ensure that awareness efforts are sustained over time by continuously innovating and adapting communication strategies to keep millets top-of-mind for consumers. Collaboration among government bodies, research institutions, agricultural sectors, and the food industry can advance millet consumption in the UAE. Joint efforts can elevate public awareness about the nutritional benefits of millets in combating metabolic syndrome, and collaborative campaigns can engage healthcare professionals to endorse millet-rich diets.
g. **Monitoring and Evaluation**—Establish a system for tracking the adoption of millet-based diets and their impact on reducing metabolic syndrome. Continuously assess the carbon footprint reduction achieved through millet-based agriculture and food systems. Adjust policies based on the outcomes of monitoring and evaluation.

h. **Incentives and Recognition**—Provide awards and recognition for individuals, businesses, and communities that make significant contributions to promoting millet-based diets and reducing metabolic syndrome.
i. **International Collaboration**—Collaborate with international organizations and other countries to share best practices and promote millet-based food systems globally. Participate in research initiatives and trade agreements related to millet and sustainable agriculture. Additionally, collaboration can drive millet cultivation in the UAE's arid environment, with experts pooling knowledge to develop drought-resistant varieties and sustainable farming techniques. Collaborate with private sector companies to invest in millet production and processing.
j. **Market Access and Distribution**—Facilitate access to markets for millet farmers by creating direct marketing channels, such as farmers' markets and cooperatives. Work with retailers to ensure the availability of millet-based products in grocery stores and restaurants. Implement pricing policies to make millet-based foods competitive with other grains. Partner with supermarkets and grocery stores to prominently display and promote millet products. Set up millet stalls at farmers' markets, fairs, and public events.

Implementing these policy recommendations will require strong government commitment, stakeholder engagement, and long-term planning. Regular assessments and adjustments will be necessary to ensure the success of the millet-based food system in reducing metabolic syndrome and carbon footprint in the UAE. Such coordinated efforts align with the nation's environmental commitments, reducing water consumption and greenhouse gas emissions associated with traditional agriculture. Food manufacturers can contribute by innovating millet-based products incorporating local flavors. In collaboration with researchers, it is possible to refine processing methods in a way that retains maximum nutrients while enhancing taste and texture at the same time. By uniting stakeholders towards a common goal, collaboration fosters improved public health and positions the UAE as a global leader in sustainable agriculture and health-conscious initiatives. The synergy between various sectors harnesses collective expertise, amplifying the impact of millet consumption on human wellness and texture.

5 Conclusion and Way Forward

Millets present a compelling solution to combat metabolic syndrome and environmental challenges at the same time. Their high nutritional value, containing fibre, vitamins, and minerals, can help reduce metabolic disorders such as obesity and diabetes. Moreover, millet cultivation offers significant environmental benefits. Thriving in arid conditions with minimal water requirements, millets align perfectly with the UAE's sustainability objectives. Their low carbon footprint and reduced water usage make them an environmentally friendly alternative to traditional crops, effectively contributing to the preservation of precious resources and climate change mitigation. As the UAE aims to establish itself as a progressive, health-conscious, and eco-friendly nation, millets emerge as a practical choice. The connection between improved public health and ecological conservation highlights the potential of millets to drive positive change on multiple fronts. Through fostering awareness, research, and collaboration, the UAE can utilize the power of millets to create a brighter, healthier, and more sustainable future for both its citizens and the planet.

(a) Develop a comprehensive policy document outlining the goals, strategies, and timelines for implementing these recommendations.
(b) Create a dedicated task force or agency responsible for overseeing the execution of millet-based food system policies.
(c) Establish regular reporting and assessment mechanisms to track progress and make necessary adjustments.
(d) Collaborate with relevant stakeholders, including farmers, nutritionists, healthcare professionals, and environmental experts, to ensure policy effectiveness.
(e) Continuously engage with the public to build support and understanding of the benefits of a millet-based food system.
(f) Evaluate the economic impact of millet cultivation and processing, including job creation and export potential.

References

Anitha, S., Botha, R., Kane-Potaka, J., Givens, D. I., Rajendran, A., Tsusaka, T. W., & Bhandari, R. K. (2021). Can millet consumption help manage hyperlipidemia and obesity?: A systematic review and meta-analysis. *Frontiers in Nutrition, 8*, 700778. https://doi.org/10.3389/FNUT.2021.700778/BIBTEX

Anitha, S., Tsusaka, T. W., Botha, R., Givens, D. I., Rajendran, A., & Bhandari, R. K. (2022). Are millets more effective in managing hyperlipidaemia and obesity than major cereal staples? A systematic review and meta-analysis. *Sustainability, 14*(11), 6659. https://doi.org/10.3390/su14116659

Belal, A. (2009). *Nutrition-related chronic diseases epidemic in UAE: Can we stand to STOP it?*

Ceasar, A. (2021). Genome-editing in millets: Current knowledge and future perspectives. *Molecular Biology Reports, 49*(1), 773–781. https://doi.org/10.1007/s11033-021-06975-w

Devkota, R., Khadka, K., Gartaula, H., Shrestha, A., Karki, S., Patel, K., & Chaudhary, P. (2016). Gender and labour efficiency in finger millet production in Nepal. In *Transforming gender and food security in the global south*. Routledge. ISBN 9781138680418.

Food security. Emirates Nature-WWF. Retrieved August 15, 2024, from https://www.emiratesnaturewwf.ae/en/our-work/ensuring-food-security.

Goron, T. L., & Raizada, M. N. (2015). Genetic diversity and genomic resources available for the small millet crops to accelerate a new green revolution. *Frontiers in Plant Science, 6*, 157.

Global Nutrition Report | Country Nutrition Profiles—Global Nutrition Report. Retrieved August 15, 2024, from https://globalnutritionreport.org/resources/nutrition-profiles/?country-search=UAE.

Gupta, S. M., Arora, S., Mirza, N., Pande, A., Lata, C., Puranik, S., Kumar, J., & Kumar, A. (2017). Finger millet: A "certain" crop for an "uncertain" future and a solution to food insecurity and hidden hunger under stressful environments. *Frontiers in Plant Science, 8*, 643. https://doi.org/10.3389/fpls.2017.00643

Gururani, K., Sood, S., Kumar, A., Joshi, D. C., Pandey, D., & Sharma, A. R. (2021). Mainstreaming Barahnaja cultivation for food and nutritional security in the Himalayan region. *Biodiversity and Conservation, 30*(3), 551–574. https://doi.org/10.1007/S10531-021-02123-9/TABLES/3

Haroun, D., Mechli, R., Sahuri, R., Alkhatib, S., Obeid, O., El Mallah, C., Wood, L., & Alsuwaidi, K. (2018). Metabolic syndrome among adolescents in Dubai, United Arab Emirates, is attributable to the high prevalence of low HDL levels: A cross-sectional study. *BMC Public Health, 18*(1). https://doi.org/10.1186/S12889-018-6215-X.

Health and climate change: country profile 2019: United Arab Emirates. Retrieved August 9, 2023, from https://www.who.int/publications/i/item/WHO-CED-PHE-EPE-19.3.2.

https://www.emiratesnaturewwf.ae/en/food-water-security.

Kane-Potaka, J., Anitha, S., Tsusaka, T. W., Botha, R., Budumuru, M., Upadhyay, S., Kumar, P., Mallesh, K., Hunasgi, R., Jalagam, A. K., & Nedumaran, S. (2021). Assessing millets and sorghum consumption behavior in urban India: A large-scale survey. *Frontiers Sustainable Food Systems, 5*, 260. https://doi.org/10.3389/fsufs.2021.680777

Mahmoud, I., & Sulaiman, N. (2021). Prevalence of metabolic syndrome and associated risk factors in the united Arab emirates: A cross-sectional population-based study. *Frontiers in Public Health, 9*, 811006. https://doi.org/10.3389/FPUBH.2021.811006

Mainstreaming millets in G20 countries through integrated behavioural and policy shifts. ORF. Retrieved October 25, 2023, from https://www.orfonline.org/research/mainstreaming-millets-in-g20-countries/.

Millets—Good for people, the environment, and farmers. Retrieved August 14, 2023, from https://unric.org/en/millets-good-for-people-the-environment-and-farmers/.

Millets show promise for dietary diversification in arid, saline environments. International Center for Biosaline Agriculture. Retrieved October 26, 2023, from https://www.biosaline.org/news/2022-05-14-13126.

Noncommunicable diseases: Risk factors. Retrieved August 14, 2023, from https://www.who.int/data/gho/data/themes/topics/topic-details/GHO/ncd-risk-factors.

Rao, A., Prabhu, D., Krishnamurthy, V., & Thimmaiah, A. K. (2019). Millets in metabolic syndrome—Time to tap the potential. *Journal of Evidence Based Medicine and Healthcare, 6*(17), 1284–1287. https://doi.org/10.18410/JEBMH/2019/266.

Saha, D., Gowda, M. V. C., Arya, L., Verma, M., & Bansal, K. C. (2016). Genetic and genomic resources of small millets. *CRC Critical Reviews in Plant Sciences, 35*, 56–79. https://doi.org/10.1080/07352689.2016.11479

Saxena, R., Vanga, S. K., Wang, J., Orsat, V., & Raghavan, V. (2018). Millets for food security in the context of climate change: A review. *Sustainability, 10*, 72228. https://doi.org/10.3390/su10072228

Thilakarathna, M. S., & Raizada, M. N. (2015). A review of nutrient management studies involving finger millet in the semi-arid tropics of Asia and Africa. *Agronomy, 5*, 262–290. https://doi.org/10.3390/agronomy5030262

The UAE climate change action strategy. Harvard Business Publishing. Retrieved August 14, 2023, from https://www.harvardbusiness.org/how-the-uae-is-committing-to-climate-change-action/.

UAE: Fossil fuel CO_2 emissions 1970–2021. Statista. Retrieved August 14, 2023, from https://www.statista.com/statistics/486080/co2-emissions-united-arab-emirates-fossil-fuel-and-industrial-purposes/.

Towards a Holistic Assessment Framework for Sustainable Household-Level Food-Energy-Water Systems

Ebrahim Karan, Mojtaba Maghrebi and Alan Sanchez

Abstract

Sustainability at the household level holds paramount importance due to its direct, tangible effects on local ecosystems and feasibility. Small-scale systems allow for a more detailed analysis of the interactions between food, energy, and water (FEW). While large centralized systems may have their own advantages in terms of economies of scale, the importance of diversifying our approach to include smaller-scale systems cannot be understated, especially in the context of building a more sustainable and resilient future. Furthermore, FEW systems at the household level often exhibit higher levels of resilience and adaptability to changing conditions, such as climate variability or market fluctuations. This study develops a holistic framework for quantitatively measuring sustainability in each of these FEW domains. Container gardening, indoor hydroponics, and backyard chickens are household-level food systems analyzed in this study based on their inputs (e.g., water, energy) and outputs (e.g., crop yields, waste generation). Active and passive solar systems, wind turbines, and geothermal are household-level energy systems analyzed in this study based on their share of renewable energy sources in the overall energy mix and their economy viability. Rainwater harvesting and greywater recycling are household-level water systems analyzed in this study based on water availability relative to demand. Integrating these measurements provides a holistic understanding of sustainability performance and guides informed decision-making for achieving more sustainable food, energy, and water systems.

Keywords

Food · Energy · Water · Sustainability · Household-level systems

E. Karan (✉) · A. Sanchez
Department of Engineering Technology, Sam Houston State University, Huntsville, TX, USA
e-mail: epk008@shsu.edu

A. Sanchez
e-mail: axs275@shsu.edu

M. Maghrebi
Department of Civil Engineering, Ferdowsi University of Mashhad, Mashhad, Iran
e-mail: mojtabamaghrebi@um.ac.ir

1 Introduction

Studying small-scale food, energy, and water (FEW) systems in terms of sustainability is crucial for a comprehensive understanding of how these interconnected systems contribute to environmental, social, and economic well-being. Such studies offer insights into the intricate dynamics that shape the resilience, efficiency, and long-term viability of communities. By studying these systems, researchers can uncover complex relationships and feedback loops that might not be as apparent in larger, more centralized systems. Furthermore, small-scale FEW systems often exhibit higher levels of resilience and adaptability to changing conditions, such as climate variability or market fluctuations (Hanes et al., 2018). Small-scale systems encourage experimentation and innovation due to their manageable size and reduced risk. By studying these systems, researchers can uncover innovative practices that have the potential to be scaled up and applied to larger contexts (Nicholls et al., 2020).

United Nations' new sustainable development agenda for 2030 has come into force since 2016, which initiates 17 sustainable development goals (SDGs) (Cai & Choi, 2020). Many of these SDGs are closely linked to small-scale systems, including goals related to zero hunger, clean water

M. Amer et al. (eds.), *Territorial Landscapes Solutions for Sustainable Energy and Food Security*,
Sustainable Landscape Planning and Natural Resources Management, https://doi.org/10.1007/978-3-031-82419-7_9

and sanitation, affordable and clean energy, and responsible consumption and production. While the importance of small-scale systems for sustainability is increasingly being recognized, academic studies capturing the viability of such systems remain relatively scarce. Large-scale projects and topics often receive more funding and attention, leaving smaller-scale systems with fewer resources for research and development. Research often focuses on cutting-edge technologies and innovations, which are more associated with larger-scale projects and industries. Furthermore, the interconnection between FEW systems may require interdisciplinary collaboration and localized investigations. Global sustainability issues might overshadow local or household-level systems in research discussions.

Studying food, energy, and water systems as a nexus, rather than in isolation, is essential because these systems are intricately interconnected. The concept of the FEW nexus recognizes that changes or disruptions in one system can have cascading effects on the others (Di Martino et al., 2023). Food production requires water and energy inputs, while energy production often requires water resources. Studying these systems in isolation can lead to suboptimal resource allocation and unintended consequences (Abdi et al., 2020). Many sustainable development goals, such as reducing CO_2, enhancing water security, or ensuring food access, are interconnected. Integrated approaches lead to more efficient achievement of these goals (Yuan & Lo, 2020).

The objective of this study is to measure the efficiency of resource use within household-level FEW systems, and assess the extent to which outputs align with resource inputs. Once the system efficiency is measured, we will be able to investigate the economic feasibility of implementing and maintaining small-scale systems, considering costs, benefits, and potential returns on investment (research goal). The next section briefly described the scope of work (FEW systems analyzed in this study) and the factors affecting the system efficiency. The development of comprehensive metrics that capture the multi-dimensional aspects of sustainability across FEW systems is explained in the third section.

2 Household-level FEW Systems

In recent years, technological advances have ushered in a new era of innovation in small-scale FEW systems designed for homes. Smart grid systems allow homeowners to monitor and optimize energy consumption, leveraging real-time data to make informed decisions (Souhe et al., 2021). Integrating renewable energy sources such as solar panels and wind turbines has become increasingly efficient, providing households with clean and decentralized energy solutions (Hoang & Nguyen, 2021). Water-saving technologies, including smart irrigation controllers and greywater recycling systems, contribute to efficient water use and conservation (Aina et al., 2023). Additionally, precision agriculture techniques and indoor hydroponic systems empower individuals to grow fresh produce year-round, utilizing minimal space and resources (Chowdhury et al., 2020). These technological breakthroughs not only enhance sustainability but also promote self-sufficiency, resilience, and reduced environmental impact at the household level.

2.1 Household-level Food Systems

Commercially available food systems designed for home use have revolutionized the way individuals interact with food production. From countertop hydroponic gardens to compact aquaponic setups, these innovative systems offer convenient and accessible solutions for cultivating fresh, nutrient-rich produce within the confines of a home environment. With automated features that regulate light, water, and nutrient delivery, these systems simplify the gardening process, making it accessible even to those without prior gardening experience (Mamatha & Namratha, 2017). These systems not only empower individuals to take charge of their food sources but also contribute to sustainability by reducing the carbon footprint associated with long-distance food transportation. Container gardening, indoor hydroponics, and backyard chickens are three household-level food systems analyzed in this study. Container gardening is a popular and versatile method of cultivating plants, typically flowers, herbs, vegetables, or small ornamental trees, in containers such as pots, planters, buckets, or other suitable containers (Torchia et al., 2023). This gardening approach is particularly well-suited for individuals with limited outdoor space, such as balconies, patios, decks, or even windowsills. In container gardening, plants are grown in soil or growing media contained within the chosen containers. These containers can vary in size, material, and style, allowing for creativity and flexibility in designing a personalized garden. Figure 1 indicates the options available to grow in containers, their container size and recommended lighting, most of these vegetables need full sun to grow, ideally between 6 and 8 h of direct light every day.

Indoor hydroponics is an innovative and sustainable agricultural practice that involves growing plants in a soilless environment, utilizing nutrient-rich water solutions to deliver essential minerals directly to the plant roots. This method offers several environmental benefits, such as reduced water consumption compared to traditional soil-based farming and the elimination of the need for harmful pesticides (Avgoustaki & Xydis, 2020). By optimizing resource usage and minimizing the impact of external

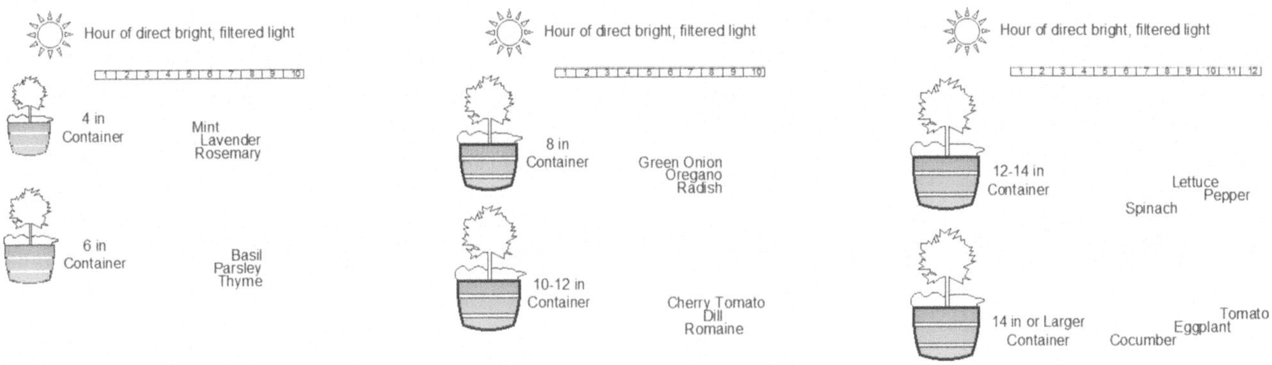

Fig. 1 Container size and lighting requirements for indoor gardening

factors like weather, indoor hydroponics can yield higher crop yields and faster growth rates. Commercially available products for indoor hydroponics encompass a wide range of systems, including nutrient delivery systems, grow lights, grow mediums, and monitoring technologies. These products cater to various scales and levels of expertise, making hydroponics accessible to both hobbyists and large-scale agricultural operations. As the demand for sustainable and locally grown produce continues to rise, indoor hydroponics stands out as a promising solution for enhancing food security and minimizing the ecological footprint of farming practices (Sathyanarayana et al., 2022). One of the popular and versatile hydroponics systems is the Nutrient Film Technique (NFT) that provides a thin film of nutrient-rich water to the plant roots, allowing them to absorb the nutrients and water directly (Rozilan et al., 2021).

Last, backyard chickens refer to the practice of keeping a small flock of chickens on residential properties, typically in suburban or urban settings, mainly for the purpose of producing eggs. It has become increasingly popular to provide fresh and locally sourced food while also enjoying the benefits of connecting with nature and practicing sustainable living. In the United States, city ordinances determine whether households can have animals such as chickens, turkeys, geese and ducks (known as fowl) on their property. According to a survey administered to a total of 9,206 pet owners in the United States, it is estimated that more than 10 million U.S. households already own backyard chickens (APPA, 2022).

2.2 Household-level Energy Systems

Household energy needs can be supplied by a combination of different sustainable sources, ranging from small wind turbines to solar photovoltaic (PV) systems, and geothermal heat pumps. The choice of a small-scale energy system depends on factors like location (amount of wind or sunlight), budget, and energy needs. Many homeowners are increasingly adopting these systems to reduce their reliance on conventional fossil fuels, lower energy costs, and contribute to a more sustainable energy future. Small wind turbines can be installed on residential properties to harness wind energy and generate electricity. They are commonly installed 30 feet (9 m) above anything within 300 feet (90 m) and especially useful in areas with consistent wind patterns (Aravindhan et al., 2023). Solar systems capture sunlight and convert it into heat (passive systems) or into electricity using solar panels (active systems). Geothermal heat pumps use the thermal energy of the ground or groundwater to provide efficient heating and cooling for homes. Residential geothermal heat pumps are closed loop systems in which the heat transfer fluid is permanently contained in a closed piping system. The space conditioning is provided by a geothermal heat pump by the use of an indoor air heat exchange coil, or indoor refrigerant-to-water heat exchanger(s), or the refrigerant circulating in pipes buried in the ground or submerged in water (Cui et al., 2019).

2.3 Household-level Water Systems

Rainwater harvesting and greywater recycling systems are environmentally friendly and cost-effective ways to collect rainwater or recycle greywater for various uses, including irrigation, flushing toilets, laundry, and even for drinking (with proper treatment). These systems are especially popular in areas with limited access to clean water or where water conservation is a priority. The major components of a rainwater harvesting system are catchment surface (usually the roof), gutters and downspouts, storage tank, a filtration system, overflow system, and a distribution mechanism. Greywater recycling systems can recycle water from bathroom sinks, shower, bathtub, laundry, and dishwashing. Greywater undergoes treatment to remove contaminants, particles, and bacteria. Treatment methods depend on the source of greywater and can include filtration, chemical treatment, and biological processes. Sand filtration is commonly used in greywater treatment systems to remove suspended solids, particles, and certain contaminants from greywater before it is reused for non-potable purposes. After large debris and contaminants are removed from the greywater, it is then directed through a sand filter bed. As greywater flows through the sand bed, suspended particles, organic matter, and some bacteria are trapped and removed. Over time, the sand filter will accumulate contaminants and particles, reducing its effectiveness. To maintain its performance, sand filters need periodic backwashing.

3 Measuring Sustainability of Household-Food Systems

The US federal intake recommendations are used in this study to quantify the energy and water requirements for the household food systems. Depending on their age and sex federal guidelines recommend that adults eat at least 2–3 cups per day of vegetables. A list of fifteen vegetables that are well-suited for container gardening is compiled with their total water and energy footprints. As can be seen in Fig. 2, an average of 19.6 L/day water is needed to meet the recommended adult's intake. For the total energy footprint, the growth stage, the type and efficiency of the lighting system used, and environmental conditions such as temperature and humidity shall be taken into account. During the germination and seedling stage, green plants require lower light intensity. LED grow lights with a total wattage of around 10–20 watts per square foot (100–200 W/m^2) are typically sufficient. LED grow lights with a total wattage of 20–40 watts per square foot (200–400 W/m^2) are often used during the vegetative growth stage. When the plants are reaching maturity and forming heads (if applicable), they benefit from higher light intensity. LED grow lights with a total wattage of 40–60 watts per square foot (400–600 W/m^2) may be necessary during this stage. An average of 6.5 kWh/day energy is needed to meet the recommended adult's intake.

Chicken gardening, while a rewarding endeavor, demands careful consideration of energy and water requirements. Raising chickens for egg production necessitates a reliable supply of clean water, with each chicken consuming approximately 1.2 L per day. To meet the needs of a household, a modest energy input of approximately 0.014 kWh per day is essential for maintaining the well-being of both the chickens and the egg-producing processes.

4 Measuring Sustainability of Household-Energy Systems

The following sustainability index (SI) defined as Eq. 1 is used in this study to the percentage of energy or water retrieved from renewable sources.

$$SI_{E\ or\ W} = S_{\text{renewables}}/D_{E\ or\ W} \tag{1}$$

D is the expected household energy (e.g., 30 kWh per day) or water demands.

The use of dollars as the unit of comparison for sustainability measurement allows us to evaluate the cost of energy supply from renewables versus conventional sources and assess the viability of renewable energy solutions in the context of their long-term economic competitiveness and their ability to drive a sustainable energy transition. Considering that the energy supply from a solar system is highly dependent on the sunlight, the results presented here are based on southeast Texas. Furthermore, the levelized cost of power is used to provide a standardized way to compare the cost of power generated from different energy

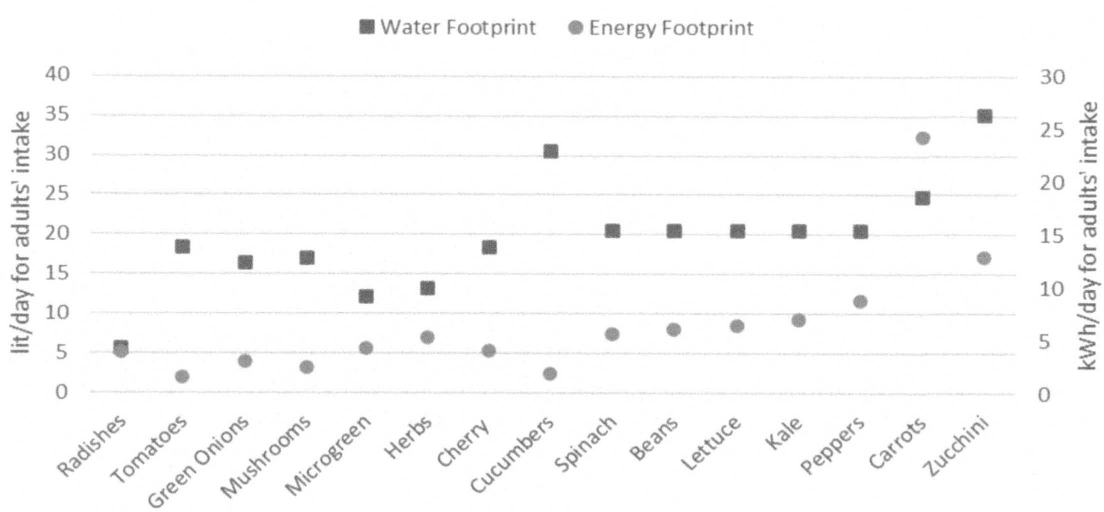

Fig. 2 Energy and water footprints for indoor gardening

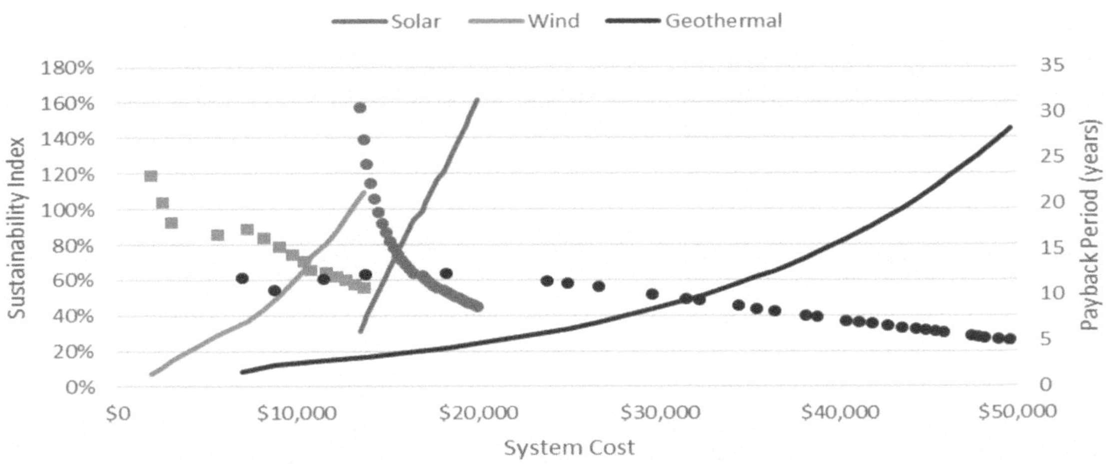

Fig. 3 Sustainability index and payback period for energy systems

sources over the entire lifetime of the system and consider the initial capital costs and operation/maintenance of such systems. We compiled a list of 46 commercially available solar systems and 51 comercially available wind turbine systems to measure the sustainability of these energy systems. The cost of geothermal heat pumps is obtained from the US Department of Energy Federal Energy Management Program. The cost of a solar system primarily depends on the number of solar panels, while the cost of a geothermal system depends on the depth of the geothermal well. The size of the wind turbine and its turbine rotor determine the cost of a wind system. Figure 3 shows the SI values for these three systems, as expected the sustainability index is directly proportional to the system's cost, meaning that a higher sustainability index corresponds to a more expensive system. The primary horzinal axis shows the normalized SI after dividing the $/kWh of the energy system by the current price of electricity in southeast Texas (13 ¢/kWh).

5 Measuring Sustainability of Household-Water Systems

The sustainability metric typically employed is the ratio of the total water recycled or harvested by these systems to the overall demand. In the context of an average household in the USA, where the daily demand is approximately 320 gallons (which converts to approximately 1211.34 L), such systems can play a pivotal role. For instance, if these systems are primarily utilized for toilet flushing (with a maximum supply of 35 gallons or approximately 132.49 L per day) and irrigation (with a maximum supply of 100 gallons or approximately 378.54 L per day), they can ideally achieve a sustainability rate of 42% by supplying 135 gallons or approximately 511.03 L of water out of the total daily demand. This metric underscores the efficiency and eco-friendliness of rainwater harvesting and greywater recycling systems in reducing the reliance on traditional water sources and mitigating the strain on freshwater reserves. The capital cost of a precipitation harvesting system can be calculated using Eq. 2:

$$\text{Cost of rainwater harvesting system} = 120 \times \sqrt{A_{\text{system}}} + (\text{System Capacity in lit}) \tag{2}$$

A_{system} is the surface area of the precipitation harvesting system (e.g., roof) in square meter.

The capital cost of a slow sand filtration system can be calculated using Eq. 3:

$$\text{Cost of greywater recycling system} = 450 \times A_{\text{tank}} + 500 \tag{3}$$

A_{tank} is the surface area of the sand filtration system in square meter.

6 Conclusion

Indoor gardening and chicken gardening emerge as sustainable solutions compared to conventional agricultural practices. They reduce the need for extensive land use, decrease transportation-related emissions, and promote the cultivation of fresh produce right at home. Although there may be an initial investment in setting up indoor gardening and chicken gardening systems, the long-term benefits in terms of sustainability and reduced environmental impact make them worthwhile choices.

For energy systems, solar power stands out as a high-sustainability option. Solar systems harness renewable energy from the sun and significantly reduce greenhouse gas emissions. While the initial investment for solar panels and installation can be substantial, the long-term benefits in terms of sustainability and reduced electricity bills are undeniable. However, for those seeking more affordable alternatives, wind turbines offer an attractive option, albeit with slightly lower sustainability. Geothermal systems present the shortest payback period, making them an attractive choice for those who prioritize a quick return on investment.

In the context of water systems, rainwater harvesting emerges as the most sustainable and cost-effective option for commercial buildings with high water consumption needs, such as toilet flushing and irrigation. Rainwater harvesting systems not only reduce the strain on municipal water supplies but also minimize water treatment costs and decrease the overall carbon footprint. When combined with a sand filtration system, the sustainability of these water systems can be boosted by up to 42%, further underlining the potential for significant long-term savings and environmental benefits.

References

Abdi, H., Shahbazitabar, M., & Mohammadi-Ivatloo, B. (2020). Food, energy and water nexus: A brief review of definitions, research, and challenges. *Inventions, 5*(4), 56.

Aina, I. V., Thiam, D. R., & Dinar, A. (2023). Economics of household preferences for water-saving technologies in urban South Africa. *Journal of Environmental Management, 339*, 117953.

APPA. (2022). *2021–2022 American Pet Products Association National Pet Owners Survey*. A. P. P. A. (APPA).

Aravindhan, N., Natarajan, M., Ponnuvel, S., & Devan, P. (2023). Recent developments and issues of small-scale wind turbines in urban residential buildings-A review. *Energy & Environment, 34*(4), 1142–1169.

Avgoustaki, D. D., & Xydis, G. (2020). How energy innovation in indoor vertical farming can improve food security, sustainability, and food safety? In *Advances in food security and sustainability* (Vol. 5, pp. 1–51). Elsevier.

Cai, Y.-J., & Choi, T.-M. (2020). A United Nations' sustainable development Goals perspective for sustainable textile and apparel supply chain management. *Transportation Research Part e: Logistics and Transportation Review, 141*, 102010.

Chowdhury, M. E., Khandakar, A., Ahmed, S., Al-Khuzaei, F., Hamdalla, J., Haque, F., Reaz, M. B. I., Al Shafei, A., & Al-Emadi, N. (2020). Design, construction and testing of iot based automated indoor vertical hydroponics farming test-bed in qatar. *Sensors, 20*(19), 5637.

Cui, Y., Zhu, J., Twaha, S., Chu, J., Bai, H., Huang, K., Chen, X., Zoras, S., & Soleimani, Z. (2019). Techno-economic assessment of the horizontal geothermal heat pump systems: A comprehensive review. *Energy Conversion and Management, 191*, 208–236.

Di Martino, M., Linke, P., & Pistikopoulos, E. N. (2023). A comprehensive classification of food-energy-water nexus optimization studies: State of the art. *Journal of Cleaner Production*, 138293.

Hanes, R. J., Gopalakrishnan, V., & Bakshi, B. R. (2018). Including nature in the food-energy-water nexus can improve sustainability across multiple ecosystem services. *Resources, Conservation and Recycling, 137*, 214–228.

Hoang, A. T., & Nguyen, X. P. (2021). Integrating renewable sources into energy system for smart city as a sagacious strategy towards clean and sustainable process. *Journal of Cleaner Production, 305*, 127161.

Mamatha, M., & Namratha, S. (2017). Design & implementation of indoor farming using automated aquaponics system. In *2017 IEEE International Conference on Smart Technologies and Management for Computing, Communication, Controls, Energy and Materials (ICSTM)*

Nicholls, E., Ely, A., Birkin, L., Basu, P., & Goulson, D. (2020). The contribution of small-scale food production in urban areas to the sustainable development goals: A review and case study. *Sustainability Science, 15*, 1585–1599.

Rozilan, M. R., Mohd Rodzi, A. S., Faiz Zubair, A., Hemdi, A. R., Deraman, R., & Md Sin, N. D. (2021). Design and fabrication of nutrient film technique (NFT) hydroponic system. In *International Conference on Mechanical Engineering Research*

Sathyanarayana, S. R., Gangadhar, W. V., Badrinath, M. G., Ravindra, R. M., & Shriramrao, A. U. (2022). Hydroponics: An intensified agriculture practice to improve food production. *Reviews in Agricultural Science, 10*, 101–114.

Souhe, F. G. Y., Mbey, C. F., Boum, A. T., & Ele, P. (2021). Forecasting of electrical energy consumption of households in a smart grid. *International Journal of Energy Economics and Policy, 11*(6), 221–233.

Torchia, D., Fresta, J., Corazza, L., & Certomà, C. (2023). New European Bauhaus for a circular economy and waste management: The lived experience of a community container garden at the University of Turin. *Sustainability, 15*(2), 914.

Yuan, M.-H., & Lo, S.-L. (2020). Ecosystem services and sustainable development: Perspectives from the food-energy-water Nexus. *Ecosystem Services, 46*, 101217.

A Comprehensive Approach to Environmental Solutions

Integrating Impact Indicators for Environmental Education in the Planning and Management of Protected Areas

Estéfany Fiorella Ramos Muñoz and Naldi Susan Carrión Puelles

Abstract

Within the global increase of protected areas (PA), local communities (LC) training involvement and engagement have become the key factors for their sustainable planning and management. This role is based on tangible achievements derived from conservation actions implementation, ecosystem services approach embracement, and territorial management improvements. However, not all LC get access to funding or support to monitor and report their progress and achievements. Thus, there is an opportunity to develop and implement managerial tools aim to improve LC' decision-making processes. Among diverse tools to promote community engagement and sustainability, environmental education (EE) has proven as crucial. EE contributes to individual empowerment in order to face environmental challenges and propose effective solutions. Moreover, its impact is not only limited to knowledge transmission or people training; its success can also be evaluated through tangible or physical improvements within the environment. Furthermore, it is possible to use social and environmental impact indicators, which would improve EE processes impact monitoring, supervising, and measuring. In this sense, through a qualitative approach, we identified indicators used on previous EE processes research and PA' management reports sent by LC to national authorities in Peru. With selected EE impact indicators, we propose a framework to integrate an educational process with PA' planning and management processes. We aim to contribute with EE programs development on LC in order to improve the crucial selection of monitoring indicators that would allow results reporting to authorities or interested parties and provides them with relevant data for decision making processes.

Keywords

Impact indicators · Environmental education · Protected areas and sustainability

E. F. Ramos Muñoz (✉) · N. S. Carrión Puelles
ESAN University, Santiago de Surco, Peru
e-mail: 15101387@ue.edu.pe

N. S. Carrión Puelles
e-mail: ncarrion@esan.edu.pe

1 Introduction

Accelerated global biodiversity loss poses a significant threat to ecosystem stability and the preservation of life on Earth (Llorente-Culebras et al., 2023; Locke et al., 2021). In response to this challenge, the creation and expansion of protected areas (PA) was promoted as a fundamental strategy for biodiversity conservation. This momentum materialized the Strategic Plan of the Convention on Biological Diversity and the Aichi Targets, which underpinned a considerable increase in the extension of the PA network worldwide, especially in megadiverse countries (Bacon et al., 2019; Geldmann et al., 2019).

Although the quantitative expansion of PA has been a significant step forward, a crucial question arises as to whether this approach is sufficient to address current conservation challenges, such as biodiversity loss (Watson et al., 2020; Woodley et al., 2019). Hence, we need to move towards a more integrated and participatory PA' management, especially by actively involving the local communities (LC) (Feng et al., 2020; UNEP-WCMC & IUCN, 2022). For example, in the specific context of Peru, a megadiverse country with priority in terms of conservation (Shanee et al., 2020), communities or individual landowners could lead PA' management under the figure of Private Conservation Areas (PCA) (Shanee et al., 2017). These types of areas were promoted by the Peruvian State as an effective strategy to promote a network approach, engaging

M. Amer et al. (eds.), *Territorial Landscapes Solutions for Sustainable Energy and Food Security*,
Sustainable Landscape Planning and Natural Resources Management, https://doi.org/10.1007/978-3-031-82419-7_10

and involving LC into the management of PA. Also, LC would be committed to recognize achievements and progress of their conservation activities (such as ecotourism, environmental education, reforestation, among others) by elaborating annual reports for the National Service of Natural Areas Protected by the State (SERNANP by its Spanish acronym) (Monteferri, 2019).

However, recognizing conservation activities' environmental and social progress and results could be limited due to the lack of monitoring and communicating tools, especially in areas under private protection (Palfrey et al., 2021). In this sense, improving PA planning and management is revealed as an important issue to achieve global conservation objectives, such as species protection and ecosystem preservation. Environmental education (EE) emerges as a fundamental conservation tool (Ardoin et al., 2020) that not only contributes to the environmental literacy of communities, but also trains people to identify and solve environmental problems (Monroe & Krasny, 2016; Zorrilla-Pujana, 2017) and encouraging their active participation in the management of their territories (Ardoin et al., 2015; Clark et al., 2020).

Regarding results or impact evaluation of EE, the approach has evolved from accounting for training participation to measuring individual intention regarding environmental actions towards the possibility of generating physical impacts on the environment (Ardoin et al., 2015; Johnson, 2013; Johnson et al., 2012; Stern et al., 2013; Thomas et al., 2018). For example, improving environmental quality or helping to achieve a conservation goal (Ardoin et al., 2020). However, a full understanding of the actual impact of EE processes and their effective integration into PA management remains a nascent area of research.

In this context, we propose a comprehensive framework that addresses the need to evaluate both the final results of EE and the intermediate processes that lead to tangible improvements in environmental conservation. This framework includes relevant environmental and social impact indicators, as well as key recommendations for effectively designing and implementing educational processes and integrating them into PA planning and management. This approach not only seeks to quantify the success of EE in PA, but also to understand in depth the underlying mechanisms that lead to significant results in biodiversity conservation and environmental sustainability.

2 Methods

The research methodology used a qualitative approach consisting of two key components: (1) a literature review to identify environmental and social impact indicators, and (2) a framework development to integrate these environmental process indicators into PA planning and management. The methodological process is illustrated in Fig. 1.

2.1 Literature Review

The literature review focused on identifying environmental and social impact indicators that can be used to evaluate EE processes impact and that can be incorporated into PA planning and management (Johnson, 2013; Zorrilla-Pujana & Rossi, 2016). The review strategy consisted of using a set of keywords to identify relevant papers in databases such as Scopus and Web of Science. Additionally, given the limited availability of reported results regarding measuring environmental changes following EE processes implementation, and considering the Peruvian PA relevance, our literature review incorporated PCA reports submitted to SERNANP. These reports function as a key primary data source for the research (Singleton & Straits, 2017).

After the evaluation of each individual study, we selected those that provided vast information for our next research process. An in-depth review of educational processes results and impacts indicators reported was undertaken. Hence, each report's evaluation included a span over the last 10 years, including those documents containing information on the educational activities carried out within PCA.

Ultimately, as this research adopts a qualitative approach, content analysis was utilized to code and classify the content within papers and reports related to environmental and social impact indicators of the EE process (Ghauri & Firth, 2009; Ghauri et al., 2020).

2.2 Framework Development

The development of the second part of this research contemplates the proposal of a framework design. For this purpose, a group of environmental and social impact indicators for EE was elaborated. This includes the indicators conceptualization and data inputs for monitoring and techniques or platforms that can be used. In order to integrate these impact indicators into the planning and management of a PA, we integrated them in stages according to the steps followed on EE development processes.

It is important to highlight that at this stage of our research we have not tested the categorized EE's environmental impact indicators. Hence, we proposed a theoretical and practice-based set of indicators and monitoring techniques identified in the literature and operational reports. Thus, we aim to present a frame of reference. Further, we will develop a case study to recognize in depth the performance of these

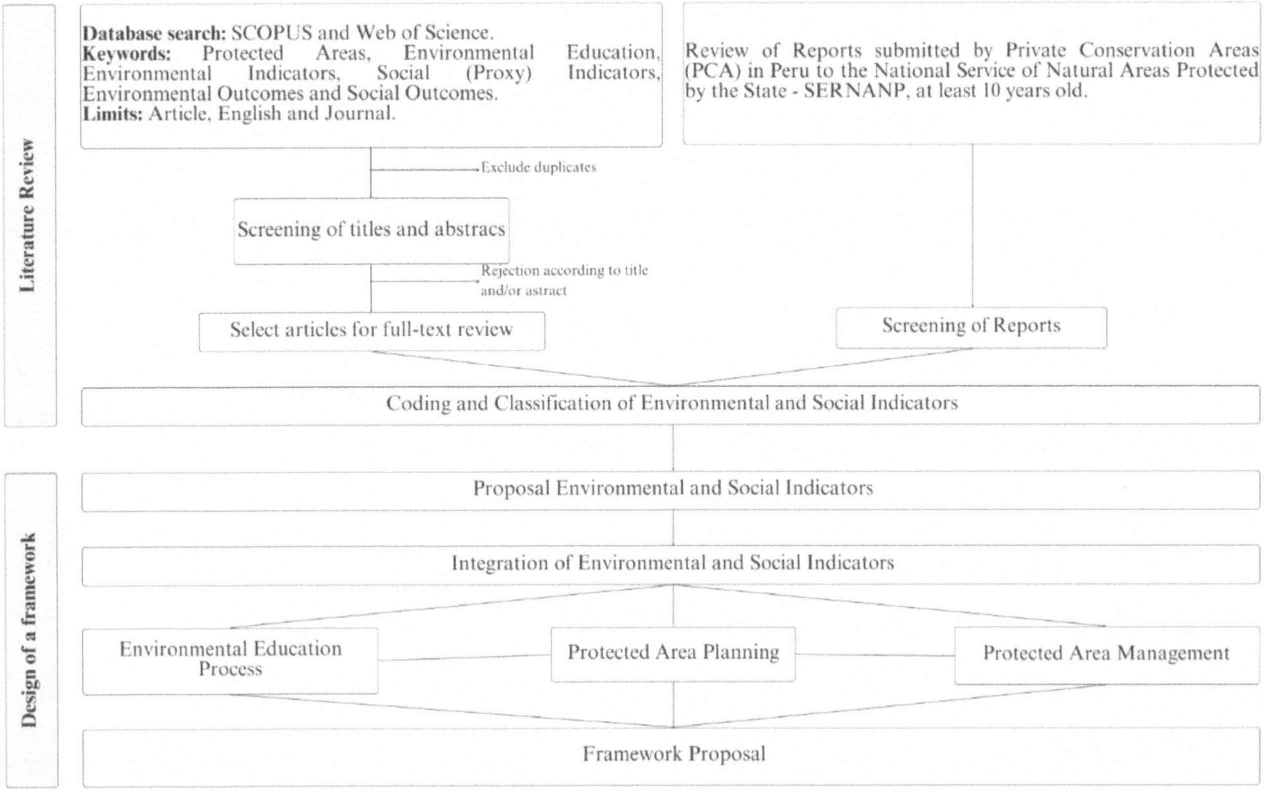

Fig. 1 Methodology

indicators, as a complementary part of this research. Thus, we will review current EE training topics throughout a longitudinal period as suggested by as Yin (2018).

3 Results

To present our results is important to explain the EE process (Fig. 2). EE programs planning and implementation processes start with topics' selection that must contain with conservation objectives or environmental local problems. The second step is to address communities or citizens training in accordance with selected topics. The third step involves implementation of hands-on conservation actions. Typically, monitoring and reporting actions cover the second and third stage, by indicating the number of people trained or their future intentions to take actions (Ardoin et al., 2020). However, we have recognized impacts that can be reported and monitored using in the first stage social impact indicators and in the second stage physical environmental indicators. Both types of indicators could be used at short- and medium-term evaluations.

At first, social impact indicators could include local teams' organization to report and control illegal

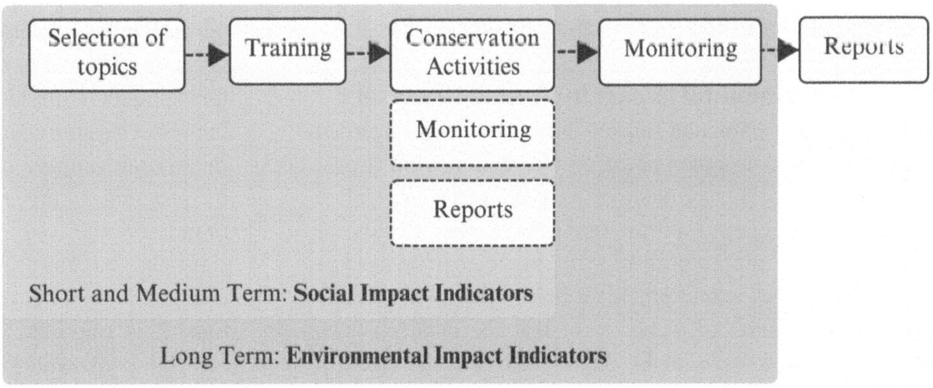

Fig. 2 Environmental education process

deforestation or wildlife hunting. Conservation activities, which underpin communities' agreements, could be reported as social indicators of citizenship engagement and organization. Thus, physical/environmental indicators could be introduced at EE's fourth step including monitoring physical environmental assets in the long term. Thus, we can use environmental impact indicators to measure environmental quality improvements or conservation progress or achievements. While these outcomes are important for decision making within LC, social and environmental outcomes can be easily communicated and reported. In the case of public and private PA, these indicators could reveal important EE outputs, which can be serve as a source of trusted information for local authorities or stakeholders.

3.1 Impact Indicators for Environmental Education Proposal

3.1.1 Social Impact Indicators Proposal

Through our research, it was learned that there are proxy indicators that can help identify the progress of EE programs before they produce a biophysical impact on the environment. Thus, proxy indicators help to identify the results of activities that are implemented as part of an educational process, considered as intermediate steps (Johnson, 2013; Johnson et al., 2012). These activities can be planting trees to reforest an area, forming committees to provide ecotourism services, or declaring a zone as a PA or establishing norms for not hunting animals.

Incorporating these types of indicators into the design of an educational process is crucial for understanding the short-term results that can be achieved through EE in a community, especially for those who manage a PA. Such results can be valuable for decision-making processes and the generation of reports that may be requested by authorities (Toomey et al., 2017; Zorrilla-Pujana & Rossi, 2016).

Finally, since the proxy indicators are related to the activities carried out by people within a PA, it was considered to call them social impact indicators. Table 1 shows a proposal for social impact indicators that can be related to EE.

3.1.2 Environmental Impact Indicators Proposal

Regarding environmental impact indicators, it was learned that they can be designed based on conventional indicators, such as water quality or biodiversity levels. However, when reviewing PA reports, it was found that signs, signals, or sightings of an important species for an area can be used as an indicator. This would provide a first look into the results of the conservation of a species or that the quality of the habitat of a species is being guaranteed and, therefore, it can be observed or signals can be found. In this sense, this type of indicator can also be used by the local population to know if the way in which they are managing their PA is working. Likewise, although it may take time (months or years) to identify these types of results from the implementation of an EE program, by using social impact indicators, the first impacts of EE can be identified, because they are interrelated.

Environmental impact indicators for EE needs continuous monitoring. For this it is important to identify what is required for monitoring, such as the development of a baseline or plan, and also techniques or tools. For example, remote sensing, data collection applications or other technological tools can be used. In this way, a robust monitoring system can be established to capture changes in the quality of the environment and the achievement of long-term conservation objectives (Kobori, 2009). Table 2 shows a proposal for environmental impact indicators.

3.2 Framework Design

The integration of social and environmental impact indicators for EE into the planning and management of PA, follow this structure (Fig. 3).

Initially, the design of an EE program should be treated as an integral element of the planning process within a PA, encompassing consideration for four key aspects:

(1) The assessment of local context to identify environmental challenges and conservation objectives of a PA. This involves consultation with community members, local authorities and conservation experts to gather insights into the unique aspects of each area, ensuring the training and educational process are tailored to address specific needs (Knight et al., 2019; Toomey, 2016; Toomey et al., 2017).
(2) Requirements for Program cover EE training topics, target audiences, instructors, and materials and resources. It is crucial to integrate conservation activities as an integral part of the program. Because designing action-oriented learning strategies, such as citizen science and project-based learning, can directly address environmental challenges, tangible physical improvements can be generated (Ardoin et al., 2020). These activities include for example: reforestation with native species, coordinating waste cleanup campaigns, or testing water quality.
(3) Incorporate impact indicators into the design of the EE program. In this way, social indicators can be used to measure the results of conservation activities in the short and medium term, and environmental impact for changes in the environment in the long term. In addition, it is important to point out that EE programs rely on conservation activities for continuity; these

Table 1 Social impact indicators proposal

Activities	Social impact indicator	Monitoring	Techniques
Conservation actions	• Vegetation cutting control rate • Vegetation burning control rate • Reforestation rate • Wildlife species hunting control rate	• Remote sensing vegetation cover monitoring Plan and field verification program • Remote sensing species distribution monitoring Plan and field verification program	Use of GEO platforms
Communal agreements	• Frequency of agreements on the PA in community assemblies	• Categorization of minutes (copies) of the communal assembly	Systematization templates
Creation of a protected area	• Frequency of agreements on the PA in community assemblies	• Categorization of minutes (copies) of the communal assembly	Systematization templates
Patrolling and/or vigilance activities	• Frequency of patrolling and/or vigilance activities	• Patrol and/or vigilance Program	Daytime line transects surveys (sightings/calls) and track and sign surveys
Creation and/or continuity of associations or committees	• Rate of committees created for the supply of ecotourism services by seniority	• Categorization of committees created according to seniority	Systematization templates

Table 2 Environmental impact indicators proposal

Environmental Impact Indicator	Estimated Time to Measure Results	Monitoring	Techniques
Variation in the population density of fauna species	3 years	• Elaboration of baseline • Study plan on population density of fauna species	• Use of GEO platforms with species distribution analysis
Rate of change of the flow of the water body	1 year	• Elaboration of baseline • Flow variation analysis plan	• Use of GEO platforms with Normalized Difference Water index (NDWI)
Water body quality	6 months	• Elaboration of baseline • Water quality analysis plan	• Conventional sampling protocols
Erosion rate of the water body margin	1 year	• Elaboration of baseline • Water body vegetation cover assessment plan with remote sensing and field verification program	• Installation of camera traps, daytime line transects surveys (sightings/calls) and track and sign surveys
Rate of wildlife habituated to human presence	3 years	• Elaboration of baseline • Study of stress hormone levels in wildlife species	• Study of the feces of the species
Rate of vegetation cover	4 years	• Elaboration of baseline • Remote sensing vegetation cover monitoring	• Use of GEO platforms with Normalized Difference Vegetation Index (NDVI)

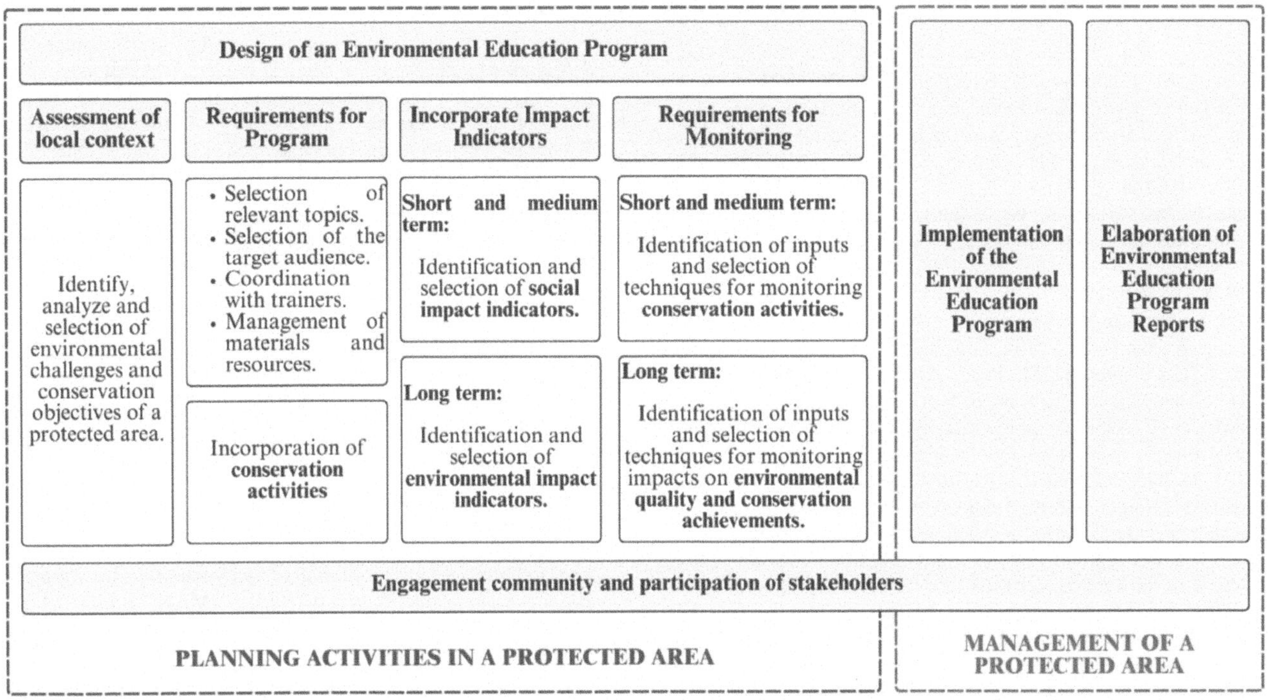

Fig. 3 Framework Design

programs can even be carried out on a permanent basis and change their theme according to the objectives of an AP or an environmental problem.

(4) The monitoring requirements show which tools should be considered for following up on social and environmental impact indicators. These requirements may include techniques such as remote sensing with the use of normalized difference vegetation index (NDVI) or normalized difference water index (NDWI), the use of camera traps or the study of a species' feces. In this sense, the proposed social and environmental impact indicators presented in this research can be used as a reference for designing indicators for a PA and incorporating them into the planning and management of the area.

After designing an EE program, it is implemented and reports are prepared based on the results achieved. These activities are carried out as part of PA management.

Finally, it's important to mention that EE promotes the engagement community and participation of stakeholders from the design of an educational program to the elaboration of results for decision making.

4 Conclusion

We recognize the fundamental relevance of EE in fostering community participation and promoting sustainability, especially when comprehensively integrated into the planning and management of PA. Therefore, when evaluating the impact of education at a social and environmental level, crucial findings are revealed that allow communities to communicate the environmental achievements of their conservation initiatives and have fundamental data for decision-making.

In this context, we emphasize the importance of incorporating social and environmental impact indicators from the initial planning phase of an EE program. Social indicators are essential to understand the progress and successes achieved by communities when they implement educational processes as part of the management of PA, such as the consolidation of community organization, participation in assemblies, and the creation of development committees of ecotourism, among other aspects. Furthermore, these indicators can be measured in the short and medium term, facilitating reporting to authorities or interested parties. Likewise, to evaluate the long-term impact of EE, environmental indicators can be used and the framework we present also establishes the estimated time frame to begin to observe tangible results in the environment.

Finally, the proposed framework is presented as a valuable starting point to reconsider EE as an essential tool for the effective management of PA, since EE empowers individuals, makes them actively involved in the administration of their territory, participating in a committed manner in decision-making. This commitment, in turn, contributes significantly to the sustainable use of resources and the long-term well-being of LC.

References

Ardoin, N. M., Biedenweg, K., & O'Connor, K. (2015). Evaluation in residential environmental education: An applied literature review of intermediary outcomes. *Applied Environmental Education and Communication, 14*(1), 43–56. https://doi.org/10.1080/1533015X.2015.1013225

Ardoin, N. M., Bowers, A. W., & Gaillard, E. (2020). Environmental education outcomes for conservation: A systematic review. *Biological Conservation, 241*, 108224. https://doi.org/10.1016/J.BIOCON.2019.108224

Bacon, E., Gannon, P., Stephen, S., Seyoum-Edjigu, E., Schmidt, M., Lang, B., Sandwith, T., Xin, J., Arora, S., Adham, K. N., Espinoza, A. J. R., Qwathekana, M., Prates, A. P. L., Shestakov, A., Cooper, D., Ervin, J., de Souza Dias, B. F., Leles, B., Attallah, M., Gidda, S. B., et al. (2019). Aichi biodiversity target 11 in the like-minded megadiverse countries. *Journal for Nature Conservation, 51*, 125723. https://doi.org/10.1016/J.JNC.2019.125723.

Clark, C. R., Heimlich, J. E., Ardoin, N. M., & Braus, J. (2020). Using a Delphi study to clarify the landscape and core outcomes in environmental education. *Environmental Education Research, 26*(3), 381–399. https://doi.org/10.1080/13504622.2020.1727859

Feng, C., Cao, M., Wang, W., Wang, H., Liu, F., Zhang, L., Du, J., Zhou, Y., Huang, W., & Li, J. (2020). Which management measures lead to better performance of China's protected areas in reducing forest loss? *Science of the Total Environment, 764*, 142895. https://doi.org/10.1016/j.scitotenv.2020.142895

Geldmann, J., Manica, A., Burgess, N. D., Coad, L., & Balmford, A. (2019). A global-level assessment of the effectiveness of protected areas at resisting anthropogenic pressures. *Proceedings of the National Academy of Sciences of the United States of America, 116*(46), 23209–23215. https://doi.org/10.1073/pnas.1908221116

Ghauri, P. N., & Firth, R. (2009). The formalization of case study research in international business *48*(1), 29–40.https://doi.org/10.1007/S12642-009-0003-1.

Ghauri, P. N., Grønhaug, K., & Strange, R. (2020). *Research methods in business studies*, 1–300. https://books.google.com/books/about/Research_Methods_in_Business_Studies.html?id=6tLKDwAAQBAJ.

Johnson, B. (2013). *Can education improve the environment? Applying the pressure-state-response environmental indicator framework to environmental education program outcomes* [Dissertation, Antioch University]. https://search.proquest.com/docview/1654132921?accountid=12528%250A; https://monash.hosted.exlibrisgroup.com/primo-explore/openurl?institution=MUA&vid=MONUI_SP&lang=en_US&?url_ver=Z39.88-2004&rft_val_fmt=info:ofi/fmt:kev:mtx:dissertation&genre=dissertations.

Johnson, B., Duffin, M., & Murphy, M. (2012). Quantifying a relationship between place-based learning and environmental quality *18*(5), 609–624. https://doi.org/10.1080/13504622.2011.640748.

Knight, A. T., Cook, C. N., Redford, K. H., Biggs, D., Romero, C., Ortega-Argueta, A., Norman, C. D., Parsons, B., Reynolds, M., Eoyang, G., & Keene, M. (2019). Improving conservation practice with principles and tools from systems thinking and evaluation. *Sustainability Science, 14*(6), 1531–1548. https://doi.org/10.1007/s11625-019-00676-x

Kobori, H. (2009). Current trends in conservation education in Japan. *Biological Conservation, 142*(9), 1950–1957. https://doi.org/10.1016/J.BIOCON.2009.04.017

Llorente-Culebras, S., Ladle, R. J., & Santos, A. M. C. (2023). Publication trends in global biodiversity research on protected areas. *Biological Conservation, 281*, 109988. https://doi.org/10.1016/J.BIOCON.2023.109988

Locke, H., Rockström, J., Bakker, P., Bapna, M., Gough, M., Lambertini, M., Morris, J., Polman, P., Carlos, M. R., Samper, C., Sanjayan, M., Zabey, E., & Zurita, P. (2021). A nature-positive world: The global goal for nature. *NaturePositive.Org*, 21.

Monroe, M. C., & Krasny, M. E. (2016). *Across the spectrum: Resources for environmental education* (3rd edn.). https://doi.org/10.1017/S1092852900019660.

Monteferri, B. (2019). Áreas de conservación privada en el Perú: avances y propuestas a 20 años de su creación. In *Sociedad Peruana de Derecho Ambiental—SPDA*. https://hdl.handle.net/20.500.12823/193.

Palfrey, R., Oldekop, J., & Holmes, G. (2021). Conservation and social outcomes of private protected areas. *Conservation Biology, 35*(4), 1098–1110. https://doi.org/10.1111/COBI.13668

Shanee, S., Shanee, N., Lock, W., & Espejo-Uribe, M. J. (2020). The development and growth of non-governmental conservation in Peru: Privately and communally protected areas. *Human Ecology, 48*(6), 681–693. https://doi.org/10.1007/s10745-020-00188-8

Shanee, S., Shanee, N., Monteferri, B., Allgas, N., Alarcon Pardo, A., & Horwich, R. H. (2017). Protected area coverage of threatened vertebrates and ecoregions in Peru: Comparison of communal, private and state reserves. *Journal of Environmental Management, 202*, 12–20. https://doi.org/10.1016/J.JENVMAN.2017.07.023

Singleton, R. A., & Straits, B. C. (2017). *Approaches to social science research*. Oxford University Press.

Stern, M. J., Powell, R. B., & Hill, D. (2013). Environmental education program evaluation in the new millennium: What do we measure and what have we learned? *Environmental Education Research, 20*(5), 581–611. https://doi.org/10.1080/13504622.2013.838749

Thomas, R. E. W., Teel, T., Bruyere, B., & Laurence, S. (2018). Metrics and outcomes of conservation education: A quarter century of lessons learned. *Environmental Education Research, 25*(2), 172–192. https://doi.org/10.1080/13504622.2018.1450849

Toomey, A. H. (2016). What happens at the gap between knowledge and practice? Spaces of encounter and misencounter between environmental scientists and local people. *Ecology and Society, 21*(2). https://doi.org/10.5751/ES-08409-210228.

Toomey, A. H., Knight, A. T., & Barlow, J. (2017). Navigating the space between research and implementation in conservation. *Conservation Letters, 10*(5), 619–625. https://doi.org/10.1111/conl.12315

UNEP-WCMC, & IUCN. (2022). *Protected planet report 2020*. https://livereport.protectedplanet.net/.

Watson, J. E. M., Keith, D. A., Strassburg, B. B. N., Venter, O., Williams, B., & Nicholson, E. (2020). Set a global target for ecosystems. *Nature, 578*(7795), 360–362; *578*, 7795 (2021). https://doi.org/10.1038/d41586-020-00446-1.

Woodley, S., Locke, H., Laffoley, D., Mackinnon, K., Sandwith, T., & Smart, J. (2019). *Una revisión de evidencias para respaldar las Metas de Conservación basadas en Áreas para el Marco Global de Biodiversidad Post-2020. 25*(November 2019).

Yin, R. K. (2018). Case study research and applications: Design and methods. *Journal of Hospitality & Tourism Research*. https://doi.org/10.1177/109634809702100108

Zorrilla Pujana, J. (2017). Integrating environmental education within the management model of National Parks of Colombia. *Environmental Education Research, 23*(5), 748. https://doi.org/10.1080/13504622.2016.1225693

Zorrilla-Pujana, J., & Rossi, S. (2016). Environmental education indicators system for protected areas management. *Ecological Indicators, 67*, 146–155. https://doi.org/10.1016/j.ecolind.2016.02.053

Beyond Western Criticism: The Benefits of the UAE's Political Neutrality and Public Diplomacy to Facilitate Dialogue at COP28

Aditi Chatterjee

Abstract

The COP28 climate summit hosted by the UAE in November 2023 had been subject to increasing criticism from policy makers and political commentators in Western countries. More than 130 European and US lawmakers published an open letter calling for COP28 president who is the head of ADNOC and also the Chairman of Masdar, to be removed, claiming that his dual role as chief executive of the country's national oil company jeopardizes the prospects of the summit. A steady stream of increasingly critical publications questions the potential of COP28 in the light of the UAE's position as an oil-producing country. This paper addresses these concerns through an assessment of the UAE's position on (and vulnerability to) climate change and environmental sustainability and a review of the key challenges to consensus building at prior COP summits. It identifies the UAE's consistent national commitment since 2005 to finding pragmatic solutions to accelerate global energy transition while reducing emissions and implementing several clean energy projects globally. Secondly, it brings into focus the highly political nature of previous COP summits where ineffective communication between national representatives driven by broader diplomatic disputes emerged as a key obstacle to agreement. This obstacle was accentuated during the pre COP28 summit climate conference in Bonn where the Russia Ukraine conflict prevented communication between key stakeholders. The UAE's position as an emerging global soft power embracing its policy of political neutrality and public diplomacy is an asset to facilitate communication across the diplomatic divide and with countries that are vulnerable to the impact of climate change.

Keywords

COP 28 · Climate change · Sustainable · Renewable · Political neutrality

1 Introduction

The COP 28 hosted by the UAE in December 2023 has been marred by allegations of prioritizing commercial interests over climate action especially by critics from the West. Despite these serious allegations the COP 28 President[1] declared in November 2023 that the COP 28 would be the most inclusive and landmark climate conference since Paris despite serious geopolitical ramifications. While he is the Chief of ADNOC, the state owned oil company of the UAE, he is also the founder and Chairman of Masdar, UAE's source of renewable energy and despite severe criticisms post COP 28 it can decidedly be said that he proved that he can deliver a deal with which most countries can live with (Meyer, 2023). The ADNOC Chief's Presidency confirmed the presence of all major fossil fuel industries at the COP 28 making it the most inclusive climate summit, alongside persuading fossil fuel industries to lower their emissions making them part of the solution. The UAE due to years of fluctuating oil revenues, globalized markets, environmental obligations and shift towards more sustainable energy markets have appreciated the need for more economic

A. Chatterjee (✉)
Assistant Professor, Abu Dhabi University, Military Program, Abu Dhabi, United Arab Emirates
e-mail: aditi.chatterjee@adu.ac.ae

[1] Head of ADNOC and Chairman of Masdar Sultan Ahmed Al Jaber presided over COP 28 held in Dubai 2023.

M. Amer et al. (eds.), *Territorial Landscapes Solutions for Sustainable Energy and Food Security*, Sustainable Landscape Planning and Natural Resources Management, https://doi.org/10.1007/978-3-031-82419-7_11

resilience and energy transition (Anshasy, 2023). In a steadily decarbonizing world, the UAE has been quick to realize the need for an improved and an economically diversified roadmap (Luciani, 2021).

2 Western Cynicism

The UAE has faced major backlash from all quarters of Western media, non-governmental organizations and even various Western governments on its decision to nominate the ADNOC Chief as the President of the COP 28 summit. Green activists have argued that the hosting of the COP 28 by a leading fossil fuel producing nation is a blatant promotion of its soft power and public diplomacy without any real action towards climate sustainability. This Western cynicism of trying to portray the UAE as the climate villain is nothing short of propagating the environmental colonialism by the Global North.

The Global North's criticism of the UAE's COP summit Presidency has been so strong that the United States Senator of Rhode Island along with his European counterparts had urged in a written complaint to the European Commission and the US President to convince UAE to replace the COP 28 President. Several other European politicians, green campaigners and civil society groups have also claimed that the UAE's choice of the COP Presidency has been an example of conflict of interest with the ideals of the COP and scandalous to say the least. For some this appointment also meant undermining the credibility of future of climate talks with the COP Summit failing to provide any stable and innovative solution to phasing out of fossil fuels (Rowling, 2023).

Though the COP 28 had a very critical role to play for climate sustainability and phasing out of fossil fuels the criticisms from the Global North to alienate the Global South even before the COP 28 had begun was hypocritical. The Western nations have been the major fossil fuel producing nations for decades and in retrospect have done and invested the bare minimum over the years towards energy transition. Similarly, when Europe's largest oil producing nation Scotland hosted the COP26 they didn't face the same criticism that was being leveled against the UAE in 2023 (Shehabl, 2023). Also, being dismissive of the UAE as a host of COP 28 on the pretext of it being an oil producing nation, the West chose to forgive other higher polluting nations of their climate obligations. The Western moral compass is so marred with duplicity that during the Russia Ukraine war, when Russian gas was cut off, the European governments did not hesitate for a moment to ally with the very same fossil fuel giant that they criticize as incapable of hosting a climate summit. Especially higher CO2 emitting nations like the United States, Australia, European Union and China needs to work more effectively towards broad based climate action instead of discrediting climate efforts by the Global South nations (Shehabl, 2023).

The deceit of the cynicism by the Global North is reprehensible and ironical at the same time because they have been the largest fossil fuel emitters with complete disregard for climate change, depriving the Global South of potential sources of development such as electricity or cooking fuel. This apathy of the West is nothing but a gimmick to deflect focus from their reluctance to decarbonize their own economies and shift the blame to the poorer nations and their needs to transition to cleaner energy. Renewable sources of energy can provide for greener solutions for the developing nations, however massive funding is essential to build the required infrastructure. The Global South lacks the finance and technology that are vital for achieving the net zero emissions that the Global North has been pushing every nation to aim for while they have been the leading producers of the green house gas emissions which caused climate change in the first place (Shehabl, 2023). While the West for so long have been efficiently able to outsource the pains of climate change to the rest of the globe, the recent flooding and wildfires across US and Europe prove that now the West cannot escape the curse of climate crisis unscathed.

This overt cynicism by the West was unable to create a setback for the COP 28 which in an unprecedented move on the inaugural day witnessed the approval of the landmark Loss and Damage Fund (LDF) which aims to help developing nations cope with the impacts of climate change. This decision was claimed to be a historic one by the COP 28 President, marking the first time a resolution was adopted on the opening day of a COP Summit (Mrinali, 2023). Thus, attempts to undermine the legitimacy of the climate summit through selective criticism because of its Presidency could not hinder it from exceeding all expectations and laying the foundation of accelerated growth and progress for the upcoming years to tackle climate change and limit temperature increases to 1.5 °C by tripling renewable energy capacity globally.

3 Popularity of the Oil Chief Amongst Developing Nations

Despite this huge controversy and strong Western criticism that has been stirred surrounding the COP28 Presidency of the ADNOC Chief, the UAE has found support in many developing nations of Asia, the strongest one being India. The Indian Prime Minister who attended the Climate Conference in Dubai affirmed that India by the year 2070 will achieve net zero emissions. UAE and India have a shared vision for embracing the energy transition in line with sustainable economic development and ensure that the global south adopts a low carbon development. The UAE has found support from other Global South nations as well, for instance, Seychelles, a small nation, off the eastern coast of Africa, which is at threat from devastating rise in sea levels as the world continues to burn planet-warming fossil fuels, lament that long-promised finance from rich countries to help it make a green transition hasn't arrived. The President of Seychelles in a recent interview enunciated, "Countries in the West promise and don't deliver, but yet they tell you, you have to keep protecting the planet," (Wells, 2023). He further voiced that while the Western nations failed to deliver it is countries like the UAE that has funded wind and solar plants in the Seychelles, adding, "I think the commitment of the UAE speaks volumes" (Wells, 2023).

However, several economists calculated that by 2030, developing nations need to invest around $2.4 trillion a year in order to de-carbonize their economies of which around $1.5 trillion will be spent on building the wind turbines and solar panels that advance the energy transition, while another $600 billion will help these economies acclimatize to the hit from a hotter planet (Hay, 2023). The remaining $300 billion will help maintain agricultural produce and protect natural habitat as the climate heats up. The finance obviously remains the grey area which needs to be funded by private investors and multilateral lenders like the World Bank. A recent survey conducted by the IEA (International Energy Agency) highlighted that the total population of developed nations of the West and their energy investment vis-a-vis that of China and other developing nations. The IEA report approximately estimates that investment in clean energy in China and the West needs to double by 2030 if the world is to limit global warming to 1.5° (International Energy Agency, 2023). For the COP 28 President the most important factor is the UAE's status as a leading oil producer and to prove its intent towards climate action. The Emirates alone, has earned over two hundred billion in net oil export revenues over the last couple of years due to high crude oil prices. UAE's intent to back the global south nations is also evident when a four and half billion-dollar scheme to deploy UAE state cash and private sector resources to help Africa decarbonize was announced by the COP 28 President in September, 2023 (Hay, 2023).

The reason why developing nations are hopeful with UAE's Presidency in COP 28 is because the net oil revenue resources in the last two years have been over two hundred billion dollars leading the ADNOC Chief to announce a $4.5 billion scheme in September 2023 to deploy UAE state cash and private sector resources to help Africa decarbonize. It definitely gives COP 28 a chance to have a more long-term impact over the previous COP sessions. For instance, in the COP 27 which was hosted in Egypt last year the Just Energy Transition Partnership (JETP) was signed by Indonesia with the US, Japan and a host of other well-off states wherein they were promised twenty billion dollars to retire coal plants, however colossal delays in receiving the fund have understandably led the Indonesian President Joko Widodo to question the commitment of the rich western world funders (Hay, 2023). Similar JETP programs in South Africa and Vietnam have also witnessed some progress but yet again at a glacial pace.

The global climate crisis have aggravated exponentially in the last couple of years which we have witnessed in a series of destructive events worldwide worrying global climate experts about the repercussions of the same. Devastating wildfires ravaged Canada, Argentina, some parts of the US and severe flooding have affected several northern and southern states of India, the most recent being the catastrophic landslides triggered by heavy rains and flood that affected Wayanad, a small southern district of India, in July 2024. In 2023 during the winter months the effects of pollution was so severe that the Indian capital New Delhi completely choked because of the thick smog that engulfed the city (Lou, 2023). Earlier this year calamitous heat waves swept across northern India, Libya, Cameroon, eastern states of the United States, Middle East, Europe and many nations in Asia resulting in deaths from heat strokes. Schools and colleges across South Asia especially in India, Bangladesh, Pakistan and Thailand have been disrupted during the summer months due to the heat waves. UAE, Brazil and Kenya even experienced major flooding of their highways and airports due to unprecedented rainfall. Countries like Myanmar and Mozambique faced destructive cyclones and Japan, Guam, the Philippines and Taiwan have been struck by tropical storms in 2023 and 2024. Such dismal climate situation calls for the rich and the developed nations to make major contributions towards climate action, something that was stressed significantly in the COP28 and pushed for through the loss and damage fund.

The UAE as directed by its President have already announced a thirty billion dollar fund named Alterra which aims to close the climate finance gap that the Global South is currently facing. The UAE also aim to generate over two hundred fifty billion dollars of investment by the year 2030 to further reduce that gap (Gokulan, 2023). The fund essentially aims to steer private markets towards climate investments and focus on transforming emerging markets and developing economies, where traditional investment has been lacking due to the higher perceived risks across those geographies (James, 2023). Various climate experts and advocacy groups have widely praised the fund's establishment by the UAE.

Developed and wealthy nations such as the United States and Japan have been criticized heavily for their meager contribution of 17.5 million dollars and 10 million dollars respectively. Climate activists have condemned these large economies for their disappointing and embarrassing contributions (Dewan, 2023). The United States which for decades has been the largest contributor of fossil fuels emissions have several times in the past objected to tying the loss and damage funding to each wealthy nation's emissions which explains their scarce contributions to the fund and their apathy to the climate cause.

UAE's commitment to renewable sources of energy is nothing new as Masdar the region's sustainability initiative to provide clean and affordable energy was launched as early as the year 2006. The UAE was the first country in the Middle East to ratify the Paris Agreement and also to announce a strategic initiative to achieve Net Zero by 2050. International Renewable Energy Agency (IRENA) which supports nations in their transitions to sustainable energy also has its headquarters in the UAE. At the COP 28 in December 2023, heads of nations decided to ripple the global renewable energy capacity by 2030 to align with IRENA's World Transition Outlook to seal the energy conversion gap and to stay on 1.5 °C. World governments have often in previous climate conferences such as Paris 2015 or Glasgow 2021 have agreed to phase out CO_2 emissions and other greenhouse emissions in pursuit of holding global temperature rise to below 2 °C and to limit rise of pre-industrial temperature levels to 1.5 °C, however the levels of commitment displayed by the nations to maintain these temperatures have been staggeringly low and in desperate need for true dedication to the climate cause (Klein et al., 2017).

4 UAE's Neutrality at COP28 and Challenges Ahead

The Emirati leadership have been successful in the creation of a global image of the UAE as the socially inclusive and politically neutral nation of the Middle East. As originally contextualised by Joseph Nye the use of force and military prowess for economic gain is too costly and dangerous for modern great powers, whereas soft power capabilities such as effective communication, national cohesion and universalistic culture helps nations attain the great power status both regionally and internationally (Nye, 1990). However, strategies and economic policies of the UAE have shifted from achieving 'traditional diversification' to focusing on broader and more climate friendly goals of 'sustainable diversification'(Anshasy, 2023). This has been incorporated by the UAE through gradual integration of long-term environmentally sustainable growth goals in its policies showcasing a clear departure from decades of oil dependency (Anshasy, 2023). The soft power resources of cultural attraction, social cohesion, public diplomacy and religious tolerance have provided the UAE with more opportunities and economic gains ranking number first in the Middle East and Arab regions and 20th globally in the Economic Opportunity Rankings. The UAE's ability to diplomatically and effectively manouvre relations with most nations have also garnered several criticisms especially from Western governments. The UAE's neutrality on the Russia Ukraine war has been viewed as a balancing act by the Emirati government because Russia is a key ally of the Middle Eastern nation. The UAE over the years have also cultivated an image of an effective mediator of international conflicts such as hosting negotiations and brokering peace between nations, as part of its soft power politics especially to impress and attract foreign investors (Boateng, 2021).

The COP28 alongside major roadblocks like funding also faced severe challenges because of the devastating geopolitical tensions in the region. For instance, the Israel-Hamas conflict was a major factor of influence in the negotiation dynamics of the Conference especially while discussions over climate financing were going on (Policy Circle Bureau, 2023). A glaring point of apprehension currently is the economic impact that the prolongation of the war could create in terms of rising oil prices which may negatively affect the ability and willingness of the developed rich nations to allocate funds for the poor climate ravaged nations (Wehrey, 2023). Russia's Ukrainian attack which began in 2022 had already cast a gloom over climate discussions held in COP27 and then seeped in to the 2023 Dubai climate summit as well. Russia was ostentatious enough to state clearly that any plans to phase out fossil fuels will not be supported by them wherein it became clear that a unanimous decision on the matter would not be feasible. Russia being the giant oil and gas producer and supplier across Europe and around the world, Russian support on the issue was of crucial importance. Interestingly as opposed to the popular Western opinion, several key players of the Global South and Africa are close ally of Russia and have not been overtly critical of Russia's assault of Ukraine.

While some diplomats from Ukraine participated in the COP28 summit, a brick roof from the war-torn Kherson region in their pavilion continued to remind everyone present at the Dubai climate summit of the collapse of the Kakhovka Dam that had completely wiped out villages and cities in Ukraine in June 2023. The devastation of the dam had led to catastrophic flooding; endangering crops in the world's breadbasket which had later led to food shortages across many regions, and additionally it had also endangered thousands of lives as it severely affected drinking water supplies, nothing short of a complete environmental chaos. While both the war torn nation continue to play the blame game as to whose fault caused the destruction of the dam, the COP28 Summit witnessed a private Ukrainian energy producer DTEK sign a memorandum of understanding with Vestas, a Danish firm, to expand its wind farm project in Mykolaiv with a whopping 450 million euro adding that the "turbines of the windmill would be spread far enough apart to survive any Russian missile attack" (Martin, 2023). The CEO of the Ukrainian DTEK enunciated that the first phase of this project was constructed for 200 million Euros ($217 million) amidst the war, with project crews spending about a third of their time in bomb shelters during the project, the geopolitical rift heavily impacting climate sustainability. The chief executive of Vestas added that "while the war has been challenging, the growing energy crisis has also led to a stronger support for renewable sources as it increases energy independence while keeping prices low" (Milne, 2022).

The primary challenge to achieve effective climate transition lies in alignment of sectoral policies with that of climate policies. The need to build a sustainable and fruitful international partnership and international interdependence between nations for long term climate sustainability has been difficult to attain (Oberthür et al., 2021). For months there were debates surrounding how Western diplomats would negotiate with Syrian President if he chose to attend the COP summit. The UAE was criticized yet again for their bold move to invite a leader who has been sanctioned by the United Nations and other European forums for the use of chemical weapons and for committing other war crimes (Hodgson, 2023). However, the political neutrality and public diplomacy of the UAE makes it the only state that has the power to maneuver all conflicting nations with careful diplomacy to work together. The UAE's motto of climate inclusivity made it the perfect host for the COP 28 because unless all players are invited to sit together and discuss, a transparent and unanimous decision alongside strong political action towards energy transition and reduction of emissions cannot be garnered. The urgency for a concerted action that represents the interests of all nations to address the climate issues is something which the UAE realizes and as a host nation of COP 28 possesses the diplomatic capability to resolve these issues in tandem with leaders of attending countries.

5 Conclusion

It is undeniable that to limit global warming to 1.5 °C a rapid reduction of carbon emissions and phase out of fossils fuels by 2035 at the latest is an absolute necessity. All countries including the UAE need to implement the phase out of fossil fuels as envisioned and agreed upon at the COP28 to avoid risking the health deterioration and lives of millions of vulnerable people. The COP28 was concluded in Dubai in December 2023 with negotiators concluding the need for tripling renewable energy capacity globally and doubling the global average annual rate of energy efficiency by 2030 along with accelerating zero and low emission technologies. The COP28 also emphasized on the need for increased focus on sustainable agriculture wherein the Emirates Declaration on Sustainable Agriculture, resilient food systems and climate action was announced with the support of over hundred and fifty countries to include emissions from agriculture and farming into their national climate action plans (World Economic Forum, 2023).

The setting up of the remarkable Loss and Damage Fund for the climate hit global south nations by incentivizing global investor to devise strategies and allocating funds of over seven hundred and twenty-six million dollars by developed nations have been one of the major achievements of this COP summit. The enhanced focus on water, oceans and coastal areas through Freshwater Challenge, an initiative joined by 37 countries to protect and restore freshwater ecosystem has been remarkable. Another significant development of the recent COP summit has been the strong momentum that was build on the issue of Climate and Health whereby new funding was pledged for health through a climate lens, because climate change causes various health risks such as malnutrition and diseases like malaria, cholera, heat strokes, respiratory illnesses like lung cancer amongst millions of people. The COP 28 also stressed on the need to accelerate the reduction of vehicle emissions through rapid development of infrastructure and deployment of zero and low emission vehicles and transition away from fossil fuels so as to achieve net zero emissions by 2050.

6 Recommendations

Though the COP28 ended on a high note, however swift and effective implementation of all pledges is required for the successful achievement of energy transition goals. While the COP 28 has been successful in mobilizing significant funding but translating these funds into tangible actions is also important to bend the emissions curve (World Economic Forum, 2023).

One of the major outcomes of the COP 28 included an agreement reached by over 130 participating nations to triple the world's renewable energy capacity and doubling of energy efficiency by the year 2030. This requires nations to considerably reduce emissions from road transport by opting for zero emissions vehicles and provision of accessible and convenient public transport. This outcome can only be achieved when all nations are able to efficiently ensure this transition, which might be tricky for developing nations facing financial challenges to invest in technologies that will make these energy transitions possible.

The operationalization of the Loss and Damage fund on the very first day of the Dubai summit has been a major takeaway for developing nations that bear the brunt of climate related damages. Institutional arrangements need to be made for the proper allocation of the funds to renewable projects alongside supervising that it is responsive to the needs of the climate vulnerable communities (Waskow, et al., 2023).

The COP 28 President also called for a future collaboration between the COP 28 COP 29 and COP 30 Presidencies to enhance international cooperation for strategizing the road map towards an effective National Climate Plans and delivering the necessary finances to incentivize action for reduction of green house gas emissions (Waskow, et al., 2023).

A remarkable outcome of the COP 28 was the recognition of the linkages between agriculture, food system and climate within the discussions. One hundred and fifty nine nations of the world signed the landmark COP 28 UAE Declaration on Sustainable Agriculture, Resilient Food Systems and Climate Action which includes food and food system in the National Climate Plans for each nation. This marks the integration of the climate resilient food system which will in future help eliminate hunger. The implementation stages however requires detailing and efficiency because though the declaration is not binding for signatories, yet the urgent need to incorporate a sustainable food system into climate goals is a significant headway.

COP 28 witnessed for the first time two hundred nations agreeing on the need to transition away from fossil fuels. To achieve this outcome it is crucial to maximize high skilled jobs linked to clean energy. It is essential especially for developed nations to expand support for training and capacity building in clean energy jobs to guarantee that this transition to renewable energy sources reaches the mass.

The COP 28 is not legally binding just like its preceding COP conferences, thus requiring global leaders to build a certain amount of momentum for the various international companies, fossil fuel producing giant corporations and civil society organizations to make strides towards an inclusive and a transparent climate sustainability and energy transition, displaying how public and private cooperation is key to positive climate action (Abu Dhabi Global Market, 2022). The global community needs to set aside all unnecessary geopolitical rifts and work closely towards making holistic climate action into a reality.

References

Abu Dhabi Global Market. (2022). UAE and the energy transition: Towards COP28. Areas of Climate Action Full Report.

Anshasy, A. A. (2023). From diversification resistance to sustainable diversification: Lessons from the UAE's public policy shift. *Management and Sustainability: An Arab Review*, 47–66.

Antwi-Boateng, O., & Alhashmi, A. A. (2021). The emergence of the United Arab Emirates as a global soft power: Current strategies and future challenges. *Economic and Political Studies*, 1–20.

Bureau, P. C. (2023). *Policy circle: Policy insights for informed governance*. Retrieved from Policy Circle: https://www.policycircle.org/environment/cop28-dubai-climate-change/.

Dewan, A. E. (2023). Nations pledge millions to new climate damage fund at COP28, US criticized for its small contribution. *CNN*.

Forum, W. E. (2023). COP 28: What did it accomplish and what's next? *World Economic Forum*.

Gokulan, D. (2023). COP28: UAE President announces $30 billion fund to bridge climate finance gap. *Gulf News*.

Hay, G. (2023). COP28's big challenge: Green cash for poor states. *Reuters*. Retrieved from Reuters: https://www.reuters.com/breakingviews/cop28s-big-challenge-green-cash-poor-states-2023-11-15/.

Hodgson, C. A. (2023). Is COP 28 destined to be a flop? *The Big Read*.

International Energy Agency. (2023). *World energy investment: Overview and key findings*. Retrieved from IEA: https://www.iea.org/reports/world-energy-investment-2023/overview-and-key-findings.

James, W. (2023). UAE president announces $30 billion fund to bridge climate finance gap. *Reuters*.

Klein, D. C. (2017). *The Paris Agreement on climate change: Analysis and commentary*. Oxford University Press.

Lou, E. (2023). *IR Insider*. Retrieved from COP28 Looms: Global climate challenges and funding dilemmas take center stage in Dubai talks. https://www.irinsider.org/science-and-environment/cop28-looms

Luciani, G. (2021). Framing the economic sustainability of oil economies. In *When can oil economies be deemed sustainable?* (pp. 9–30). Palgrave Macmillan, Singapore.

Martin, T. (2023). Vestas and DTEK sign MOU on Ukraine's largest wind energy project. *Energy Global*.

Meyer, C. (2023). COP28—A step in the right direction. *Arab News*.

Milne, R. (2022). Vestas warns war in Ukraine will add to wind industry slowdown. *Financial Times*.

Mrinali, K. (2023). COP28 opens in Dubai with landmark loss and damage fund, setting an ambitious agenda for climate talks. *Frontline: The Hindu*.

Nye, J. (1990). Soft power. *Foreign Policy*, 153–171.

Oberthür, S. T. (2021). Introduction. In S. T. Oberthür (Ed.), *European foreign policy in a decarbonising world: Challenges and opportunities* (1st ed., pp. 1–22). Routledge.

Rowling, M. (2023). Backlash as UAE oil boss picked to lead COP28 climate summit, Context: Thomson Reuters Foundation. Retrieved from Context: https://www.context.news/net-zero/backlash-as-uae-oil-boss-picked-to-lead-cop28-climate-summit.

Shehabl, M. (2023). The global north is alienating the global south before COP28 even starts. *Foreign Policy*.

Waskow, D., et.al. (2023). Unpacking COP28: Key outcomes from the Dubai climate talks, and what comes next, commentary. World Resource Institute.

Wehrey, F. (2023). *How the Israel-Gaza War could disrupt the Middle East's climate progress*. Retrieved from Carnegie Endowment For International Peace: https://carnegieendowment.org/2023/10/18/how-israel-gaza-war-could-disrupt-middle-east-s-climate-progress-pub-90796.

Wells, E. (2023). *Despite criticisms from the West, the UAE's oil chief is proving popular in the developing world*. Retrieved from CNN: https://edition.cnn.com/2023/11/24/climate/cop28-climate-uae-al-jaber/index.html.

Collaborative Planning of AI Chatbots in Online Health Communities: Sociomateriality and Sustainability Perspectives

Alain Osta, Angelika Kokkinaki and Charbel Chedrawi

Abstract

Objective: This research explores the intention of using health chatbots in Online Health Communities from sociomaterial and sustainability perspectives. The study examines the influence of variables and moderators like Performance Expectancy, Effort Expectancy, Social Influence, Facilitating Conditions, Fear of Technological Advancements, and Patient acceptance and Trust on users' intentions and their actual or potential use of these communities. The research acknowledges the interplay between materiality and the social environment in shaping technology usage while also considering sustainability within this context. Materials and Methods: A quantitative methodological approach was used to investigate users' behavior and intentions toward AI conversational agents/chatbots in OHCs. The UTAUT model was selected because it is well-established for assessing technology acceptance and usage behavior. It was extended to include additional variables relevant to the context of OHCs and AI technologies. Data collection was conducted via an online survey, which was distributed across various online platforms and networks to reach a diverse audience. The survey received 443 complete responses from participants spanning 62 different countries, providing a rich and diverse dataset for analysis.

Conclusion: Our current focus on sustainability variables is being expanded to include several new aspects. The new variables pertain to the cost-effectiveness of AI implementation in healthcare, Patient Outcomes, User experience, and long-term viability to help us understand how AI technologies impact aspects like diagnosis accuracy, medical error reduction, and treatment plans.

Keywords

AI chatbots · Online health communities · Sustainability · Sociomateriality · Collaborative planning

1 Introduction

Research indicates that the healthcare sector benefits from a symbiotic relationship fostered by Artificial Intelligence (AI) technology, which in turn promotes sustainability (Bender, 2019; Nguyen & Wang, 2021). It has been demonstrated that integrating AI-driven solutions optimizes a number of healthcare-related aspects such as treatment methods, operations, and resource use. The efficient and environmentally friendly delivery of healthcare services is improved as a result of this optimization (Hesamian et al., 2019; Mamoshina et al., 2016).

In general, AI increases resources management efficiency, which strengthens resilience in the healthcare sector. This includes minimizing energy consumption, lowering carbon emissions and conserving energy (Badidi, 2022; Hsu et al., 2021; Richie, 2022; Sood et al., 2022). AI also has the ability to improve access to healthcare services by

A. Osta (✉)
Graduate School, University of Nicosia, Nicosia, Cyprus
e-mail: osta@uls.edu.lb

A. Osta
Université La Sagesse, Forn El Chebbak, Lebanon

A. Kokkinaki · C. Chedrawi
School of Business, University of Nicosia, Nicosia, Cyprus
e-mail: kokkinaki.a@unic.ac.cy

C. Chedrawi
e-mail: charbel.chedrawi@usj.edu.lb; chedrawi.c@unic.ac.cy

C. Chedrawi
Business and Management Faculty, Saint Joseph University, Beirut, Lebanon

M. Amer et al. (eds.), *Territorial Landscapes Solutions for Sustainable Energy and Food Security*, Sustainable Landscape Planning and Natural Resources Management, https://doi.org/10.1007/978-3-031-82419-7_12

providing healthcare support to remote regions (Hadley et al., 2020).

The United Nations (UN) Member States have adopted the UN's 2030 Agenda for Sustainable Development, a global initiative that is structured around 17 Sustainable Development Goals (SDGs) (UN, 2015). These SDGs call for a collaborative global partnership and this study explores three pertinent UN SDGs; SDG3, focusing on health and well-being for all; SDG9, aiming to improve industry, innovation, and infrastructure, and SDG11, dedicated to sustainable cities and communities (UN, 2015).

In the dynamic landscape of healthcare, the accelerated adoption of innovative technology and investment in the continued development of digital health infrastructure is essential for its sustainability (Al-Kahtani et al., 2022). Technological developments have completely changed how patients may obtain high-quality medical care (Tortorella et al., 2022). However, it is critical to recognize the possible negative consequences of digital culture on healthcare, especially when it comes to digital security. Privacy and personal data issues are among them (Burton-Jones et al., 2020; Thomaz et al., 2020). Digital/virtual assistants, as well as chatbots, are types of conversational agents that have recently drawn new attention. AI conversational agents are technologies that provide natural language user interactions. AI chatbots, on the other hand, usually facilitate text-based conversations or clickable replies that mimic instant messaging applications (Koulouri et al., 2022).

Research has shown that AI chatbots may be useful for providing healthcare support, especially in Online Health Communities (OHCs) (Sweeney et al., 2021). Although the use of health chatbots is increasing, it is still unclear what aspects affect users' intentions to keep using these tools (De Barcelos Silva, 2020; Zhu et al., 2022). The intention and/or actual use of health chatbots may also be influenced by factors other than the ones included in the Unified Theory of Acceptance and Use of Technology (UTAUT) model, which include Performance Expectancy, Effort Expectancy, Social Influence, and Facilitating Conditions (Osta et al., 2022; Venkatesh et al., 2003). These variables include Fear of Technological Advancements (FOTA) and trust, as well as moderators like technology experience, chatbot experience, and geographical zone.

The intention of users to use chatbots in OHCs (viewed as the "social" aspect) and AI (often seen as the "material" aspect) are intertwined, though not immediately apparent. The concept of sociomateriality is a great fit for this relationship because it describes the constitutive entanglement between the social and material worlds as an "inherent inseparability." (Orlikowski & Scott, 2008).

This research aims to explore the motivations behind the intention of using health chatbots within Online Health Communities (OHCs) through the dual lenses of sociomateriality and sustainability perspectives. Employing a quantitative research methodology, the study seeks to examine the influence of various variables and moderators on users' intentions and their practical or potential engagement with these communities. A dataset comprising 443 complete responses collected from individuals representing sixty-two different countries, serves as the foundation for this investigation. The study seeks to examine the influence of variables and moderators like Performance Expectancy (PE), Effort Expectancy (EE), Social Influence (SI), Facilitating Conditions (FC), Fear of Technological Advancements (FOTA), and Patient acceptance and Trust on users' intentions and their actual or potential use of these communities. For instance, PE concerning interoperability involves evaluating the efficacy of AI solutions in integrating with existing healthcare systems and technologies, streamlining operations, and facilitating seamless information exchange. The research acknowledges the interplay between materiality and the social environment in shaping technology usage while also considering sustainability within this context.

To achieve this objective, this study will review the relevant literature on AI chatbots and OHCs, as well as sociomateriality, sustainability, and OHCs. Additionally, the proposed model will be presented based on the UTAUT framework, including hypotheses to be tested. The methodology used will be briefly explained, followed by a proper discussion of the findings. Finally, the study will conclude with its contributions, managerial implications, and limitations.

2 Background and Significance

2.1 AI Chatbots and OHCs: A Sociomaterial and Sustainability Approach

Chatbot mimic human interactions by solving problems, making decisions, going about daily tasks, and responding to users quickly. Developers and programmers utilize machine learning and artificial intelligence to create these chatterbots, which may interact with users through text messaging, voice calls, and voice commands. AI chatbots provide numerous avenues for illustrating a shift in behavior for disease prevention and widely endorsing health-related behaviors (Chakraborty et al., 2022). However, several questions in relation to the intentions to use health-related AI conversational agents are not fully exploited in the literature (Osta et al., 2022).

According to Abd-Alrazaq et al. (2020) and Calvaresi et al. (2019), the use of chatbots can incorporate behavior intervention into a regular clinical setting and prevent any additional burden on the current healthcare providers (low intensity services, provider-recommendation services, suggestions to the healthcare providers, integration into existing services or programs as a "support tool"). More interaction or relational skills, sympathetic dialogues, and a feeling of personal connection with users through sympathetic replies are required (Milne-Ives et al., 2020). For a long-lasting and interesting intervention, it is therefore essential that the AI-chatbots build a suitable rapport or relationship with the user through personalized and empathic interactions.

In this regard, Online Health Communities (OHCs) offer opportunities to patients, friends, and family to post and explain their concerns, ask questions, receive feedback, or share their experiences; they are interactive online health platforms that integrate functions such as seeking medical advice, exchanging health information, getting access to health advice, social support, and emotional connection (Osta et al., 2022). Users/patients can choose a physician to interact with directly to protect their privacy (Milne-Ives et al., 2020).

OHCs have different disadvantages as an Internet network, not least among those being the fact that data provided by users may not be accurate (Sillence, 2013). Furthermore, patients cannot be sure that they are getting advice from a real physician, even though physicians must present their certifications. Also, physicians cannot assess the condition of their patients in an accurate way without face-to-face meetings, which can minimize mistakes and increase patient satisfaction.

Within this context, sociomateriality represents technical and social aspects, and particularly their intertwining, which lead to routine behaviors (Cecez-Kecmanovic et al., 2014; Leonardi, 2013). The various contributions made to the field of sociomateriality lead to the recognition that there is a relationship that can be established between the level of sociomateriality and the AI system's performance. The necessity to comprehend the possible cooperation of humans and machines and its consequences for organizations is becoming increasingly apparent among academics and professionals (Chedrawi & Haddad, 2022). According to the sociomateriality perspective, an online community is the outcome of how technology and human relations are intertwined into the community's structure (Cecez-Kecmanovic et al., 2014; Leonardi, 2013).

This sociomaterial concept is alerting technology experts about the need to create AI applications that serve humanity instead of demolishing it, to use AI as a tool to produce social contributions and enhance the progress of societies. For instance, speech recognition systems assist doctors and physicians in their daily routine in hospitals. Because of the relationship between technology, organizations, and community members, scholars have pointed out that variables that promote and inhibit online communities are interconnected (Abedin & Babar, 2018; Park et al., 2019).

One example of how social and technical elements might coexist in a single identity is found in online communities. A lively online community can be created with the help of a strong technical platform, but interpersonal relations and active participation from members are also essential for the community to grow and take shape (Faraj et al., 2016). By taking into account the role of technology in addition to that of institutional and individual human actors, sociomateriality helps describe dynamic interactions in the digital world (Nambisan et al., 2017).

Richie (2022) and Hadley et al. (2020) have highlighted the broad applicability of AI in the field of sustainability, going beyond metropolitan areas to offer healthcare help in distant places, particularly in times of disaster. AI technology is able to efficiently handle large amounts of healthcare data, optimizes workflows, and identify complex patterns (Forbes, 2022). It is essential to strengthen the healthcare system's sustainability and resilience. Consequently, AI helps to preserve more lives by reducing energy consumption, cutting carbon emissions, conserving energy, mitigating loss, and reducing healthcare resource allocation (Badidi, 2022; Hsu et al., 2021; Jung et al., 2018; Richie, 2022; Sood et al., 2022).

According to Ahad et al. (2020), AI technology is distinctive among the technologies that enable sustainability. As stated by Muhammad et al. (2021), building sustainable cities is crucial since they immediately improve a number of health-related factors. Severe pollution-affected cities, especially those with high population densities, negatively impact people's health and raise the risk of chronic illnesses (Alahmari et al., 2022). Sustainability can be examined from a variety of angles and perspectives. Establishing a healthcare system that serves patients, doctors, government agencies, and healthcare organizations is one way to achieve sustainability from a societal standpoint.

Less focus has been placed on the results of this social support, despite an extensive body of literature on the factors that encourage knowledge sharing and social support in Online Health Communities (OHCs) (e.g., Yan et al., 2016; Zhang et al., 2017). Although past research has recognized the obvious and noteworthy benefits of OHCs, such as informational and emotional support, more study is required to fully understand their long-term advantages (Zhang et al., 2018). In the end, OHCs' sustainability and value are mostly determined by the long-term advantages they provide to users.

The use of AI to further Sustainable Development Goals (SDGs), like lowering CO2 emissions or improving horticultural products with machine learning is already starting to impact corporate sustainability, despite these technologies being in their early stages (Riffle, 2017).

The integration of AI into corporate decision-making involves two key areas: AI for sustainability and AI sustainability. In the first, companies adopting AI are expected to fully commit to ethical and responsible AI practices. Since there is currently no legal framework governing AI design and use, private standardization may arise, as voluntary adaptations can pave the way for future legal regulations (Darnault et al., 2019).

2.2 The Impact of AI Agents/Chatbots on OHCs' Users: The Proposed Model

The present study investigates the influence of user intention and behavior in Online Health Communities (OHCs) through the utilization of the widely employed Unified Theory of Acceptance and Use of Technology (UTAUT) model (Osta et al., 2022; Venkatesh et al., 2003). The fundamental constructs of the UTAUT model form the basis of the conceptual framework, which aims to scrutinize both the intention to use and the actual usage behavior. The model incorporates four primary constructs: Performance Expectancy (PE), Effort Expectancy (EE), Social Influence (SI), and Facilitating Conditions (FC), which serve as direct determinants of Behavioral Intention (BI) and Use Behavior (UB) (Venkatesh et al., 2003).

Prior studies have indicated the necessity of conducting a more thorough investigation into the part that emotions and fear play in the adoption of new technologies (Steinert & Roeser, 2020). Researchers have suggested that users with limited literacy may feel uncomfortable about unfamiliar technology, fearing that its use could reveal their cognitive limitations. As a result, they might decide to completely avoid using technology in order to reduce their stress (Di Giacomo et al., 2020; Reinares-Lara et al., 2018; Viswanathan & Sreekumar, 2019; Winter et al., 2020). In addition to these factors, other moderators such as geographical zones, experience in chatbots, technology experience, as well as trust and Fear of Technology Acceptance (FOTA) need to be included in the UTAUT model and tested in different contexts in future studies (Venkatesh et al., 2003).

The current study aims to investigate the impact of the proposed extensions on the acceptance level of AI agents/chatbots in OHCs, utilizing the UTAUT model. The UTAUT model was chosen due to its comprehensive framework that accounts for users' intentions to adopt new technologies. In the extended UTAUT model, the interrelations between the two additional constructs (FOTA and Trust) and the proposed moderators (Experience in Chatbots, Experience in Technology, and Geographical zones) are illustrated in the following Fig. 1.

Regarding the above-mentioned variables and their hypotheses, please refer to Appendix 1.

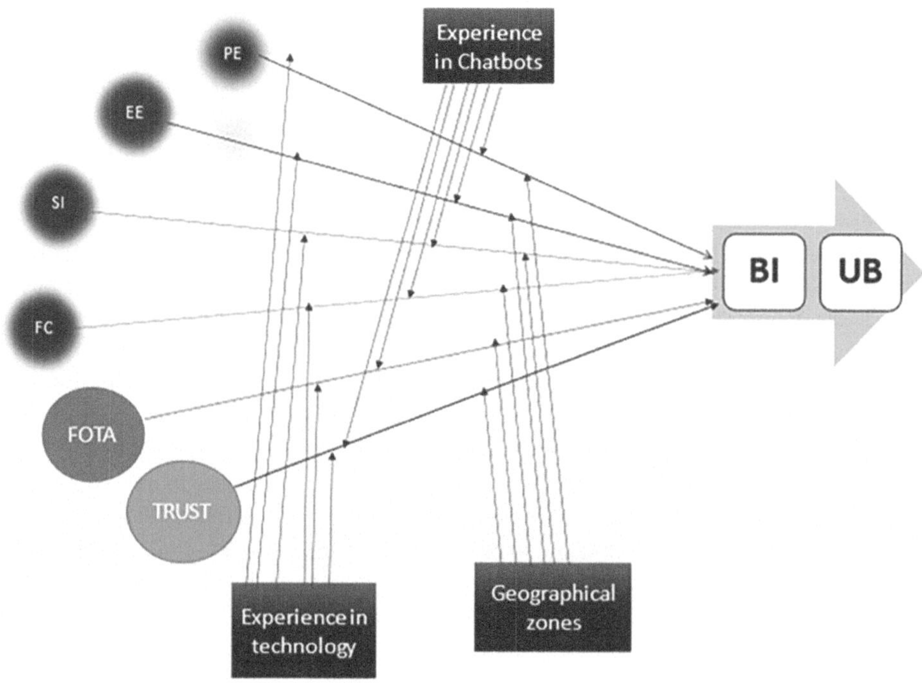

Fig. 1 Proposed extended UTAUT model

3 Materials and Methods

Upon the examination of theoretical perspectives and the formulation of hypotheses, a survey was conducted to gather data and evaluate the validity of the hypotheses (Osta et al., 2022). It has been observed that positivistic approaches are widely utilized in IS studies (Orlikowski & Baroudi, 1991), and hypothetico-deductive methods are commonly employed in quantitative studies in IS (Walsh, 2015). Bivariate analysis, specifically student t-test and ANOVA test, was employed to examine the correlation between UTAUT scores (PE, EE, SI, FC, Trust, FOTA, BI, and UB) and moderators (Experience in technology, Experience in chatbots, and Geographical zones). Linear regression was also utilized to assess the relationship between UTAUT scores and BI, as well as between BI and UB. Additionally, multivariate analysis was employed to evaluate factors influencing BI. Statistical significance was set at 5% (p-value$<$0.05).

The analysis was carried out on a convenience sample of 443 participants (members of OHCs from 62 countries) between March and May 2021 (Osta et al., 2022). The results of the study will be presented and discussed with respect to the hypothesized relationships between the primary and extended constructs of the UTAUT model and BI and UB. Furthermore, the extended moderators proposed by Venkatesh will also be examined. The participants were equally distributed by gender, with 49% females and 46% males (Appendix 2). Among the 443 participants, 57.8% had limited experience in the use of chatbots and/or online communities, and only 3.6% considered themselves experts in the use of chatbots in OHCs (Appendix 3). This low level of experience highlights the need for further research on this topic. The UTAUT variables were assessed, and all six variables were supported in the multivariate analysis. The results were consistent with most of the hypothesized interrelations with UTAUT variables (H1, H2, H3, H4, H5, H6, and H7). Linear regression indicated a positive correlation between the variables of the conceptual model and the BI to use chatbots in OHCs (PE: $p=0.000$, EE: $p=0.001$, SI: $p=0.000$, FC: $p=0.001$ and Trust: $p=0.001$). However, BI was not correlated with FOTA ($p=0.077$) (Appendix 4). These variables have been deemed crucial in determining the acceptance of technology in the healthcare context, as indicated by several research studies.

4 Results and Discussion

The present study investigated the interrelations between AI conversational agents/chatbots and OHCs, adopting sociomateriality and sustainability perspectives to explore users' Behavioral Intention. The study further examined the impact of proposed UTAUT extensions, namely Fear of Technological Advancements and Trust, on participants' Behavioral Intention and Usage Behavior towards AI conversational agents/chatbots in OHCs. The study results revealed that these extensions had a significant effect on the Behavioral Intention and Usage Behavior of participants towards AI conversational agents/chatbots in OHCs.

4.1 Correlation Between the Two Proposed Variables (Trust and FOTA) and BI

In the bivariate analysis, a statistically non-significant correlation was observed between Fear of Technological Advancements (FOTA) and Behavioral Intention (BI) ($p=0.077$). This finding contradicts the results of the literature review which emphasized the existence of technological anxiety (Guo et al., 2013) or fear of using technology (Viswanathan & Sreekumar, 2019). Technology anxiety refers to the negative emotional response that leads to users' reluctance to embrace new technology (Meuter et al., 2003). Negative anticipated emotions, including discomfort, stress, anxiety, disgust, or fear of innovation, serve as barriers to achieving desired outcomes (Bettiga & Lamberti, 2018; Tuccillo et al., 2020). These emotions also operate as obstacles to intention (Table 1).

In our population, the FOTA score demonstrated validity with a Cronbach's Alpha value of 0.852. The results revealed that participants highly disagreed with the idea of being concerned about the fast-paced advances in information technology. Additionally, they expressed comfort with technological advancements, with 88% of participants scoring within the "very good" range, indicating a lack of fear of new technologies. This finding is significant and highlights the potential for further development of chatbots in OHCs. Examining FOTA is crucial as it represents individuals' mental resistance and avoidance towards new technologies, including the use of computers, mobile apps, and other technological innovations. With almost 88% of participants showing no resistance, fear, or anxiety in using chatbots in OHCs, this underscores the favorable disposition towards the use of such technologies.

Table 1 Correlation between FOTA and Trust scores and BI

FOTA	Pearson Correlation (r)	0.084
	P-Value	0.077*
	N	443
Trust	Pearson Correlation (r)	0.632
	P-Value	<0.001*
	N	443

However, it is interesting to mention that the p. value of FOTA **in the Linear Regression for the factors affecting BI in the European population is 0.041**. Following the strength of correlation (0.611) in Europe (Appendix 5), it will be interesting to study the southern and northern European population in future studies.

The factor of Fear of Technological Advancements (FOTA) is considered to have a significant impact on Behavioral Intention (BI) (Niehaves & Plattfaut, 2014). Previous studies indicates that FOTA/anxiety negatively influences on BI (Maican et al., 2019), and its ability to predict Usage Behavior (UB) has also been established (Khechine & Lakhal, 2018). Despite existing literature showing a negative correlation between user anxiety and fear and their attitude toward adopting a new system (Compeau & Higgins, 1995; Venkatesh & Davis, 2000), our study found that FOTA was not significantly correlated with BI ($p = 0.271$) in our study population. Several factors may explain this finding. Firstly, our participants were selected from technology users, as the questionnaire was distributed to members of Online Health Communities (OHCs) on Facebook, LinkedIn, and the health website "Askapatient.com." Future studies should include participants without access to online platforms to measure their level of fear. Moreover, research has demonstrated that older users' resistance to change does not hinder their use of a system if they believe it improves their performance (PE). Additionally, users' fear of the system does not negatively affect its use, unless they perceive it as difficult to adopt (EE) (Guo et al., 2013).

Through Pearson's correlation, the findings showed that the Trust variable has a positive correlation with the BI to use chatbots in OHCs. Moreover, through linear regression **for the factors affecting BI in the Healthcare population and in the Business population,** the findings showed that Trust has a positive correlation with the BI to use chatbots in OHCs ($p = 0.001$; $p = 0.000$).

The literature has shown that adoption decisions, especially for AI medical devices, are significantly influenced by trust in AI-based solutions (Lee et al., 2003). The autonomy of AI applications also affects individuals' perception of trustworthiness. The development of a reliable environment in the healthcare industry and eHealth services is necessary for the successful implementation of technological tools (Arfi et al., 2021). In the case of the healthcare setting, trust is no longer linked to risk, and research indicates that it significantly influences BI's choice of using a technological system (Arfi et al., 2021). The results of the present study indicate that almost all participants agree that chatbots deployed in OHCs may meet their informational demands.

A statistically significant and positive correlation was found between Trust and BI ($p < 0.001$; $r = 0.632$). Trust issues are linked to ethical concerns about AI discrimination, human behavior and AI, human value judgment, machine compatibility, and moral dilemmas (Dwivedi et al., 2021). Therefore, one of the biggest obstacles to the effective integration of AI in healthcare procedures is winning over the public's trust in embracing AI healthcare (Whittlestone et al., 2019). In the present study, Trust was found to have a significant correlation with BI through Pearson's correlation ($p < 0.001$). The EE and perceived benefit were found to have a direct relationship with individuals' trust that the system is meeting their specific needs (Venkatesh et al., 2003). In general, the strongest variables in the study were PE, EE, FC, and Trust, as they appeared to have a significant correlation with BI in the different tests.

Several factors negatively impact trust in using chatbots in OHCs. Privacy concerns are always a primary issue for participants in online platforms, and bad experiences in online communities such as Facebook and WhatsApp and continuous debates around privacy issues increase the need for trust in using chatbots in OHCs. The importance and relevance of sociomateriality and the concept of humanized AI are discussed in this context. Although the relationship between trust and transparency in the context of AI conversational agents/chatbots is yet unclear, the issue of trust must always be considered when building and using AI conversational agents/chatbots systems. Understanding the distinction between social and material agency through the concept of sociomateriality can help improve the interaction between AI conversational agents/chatbots and users' intentions in OHCs. Collaboration with human-AI agents/chatbots can improve these online platforms and develop trust in AI systems' cooperative approach to human partners and their environment, particularly because patients frequently have privacy concerns and lack trust when working with AI conversational agents.

4.2 Correlation Between the UTAUT Scores and the Proposed Moderators

The correlation between UTAUT scores and the proposed moderators (experience in technology, experience in chatbots, and geographical zones) is shown in the Table 2.

The adoption of chatbots in online health communities (OHCs) is positively associated with the user's level of experience with technology, particularly in emerging markets. Linear regression analysis was performed to investigate the relationship between the attitude of individuals towards the use of chatbots in OHCs and their experience with the technology. The results showed that individuals who have limited experience with technology and chatbots, or are progressing in their use, are more likely to adopt the technology, resulting in a higher BI towards using chatbots in OHCs.

Table 2 Moderators influencing BI and UB: a bivariate analysis

Moderators	Bivariate analysis	Appendix
Experience in technology	EE, FC, and FOTA were statistically correlated No statistically significant correlation exists between the PE, SI, TRUST, BI and UB	6
Experience in chatbots	PE, EE, SI, and FOTA were statistically correlated Participants who are progressing in the use of Chatbots in OHCs had the highest PE, EE, SI, and FOTA BI and UB were statistically correlated	7
Geographical zones	PE, SI, FC, and Trust were statistically correlated. Participants living in countries other than Europe, North America, and the Arab countries had the highest PE mean BI was statistically correlated No statistically significant correlation exists between the EE, FOTA, and UB	8

Table 3 Linear regression test between UB and BI

Model summary

Model	R	R square	Adjusted R square	Std. error of the estimate
1	0.514[a]	0.264	0.262	2.992

a. Predictors: (Constant), BI

Model		Unstandardized coefficients		Standardized coefficients	T	P-Value
		B	Std. error	Beta		
1	(Constant)	5.675	0.435		13.046	0.000
	BI	0.289	0.023	0.514	12.570	0.000

a. Dependent Variable: UB

Furthermore, the regression analysis indicated a statistically significant correlation between UB and BI ($p=0.000$). The model established that UB is expected to increase significantly with the increase of BI, with a correlation strength of 0.264. These findings demonstrate the importance of the attitude towards using chatbots in OHCs, as it significantly influences the perception of benefits associated with the technology, and ultimately the likelihood of adoption. The results suggest that improving the user's experience with the technology and building a positive attitude towards chatbots may increase the adoption of the technology and positively impact UB (Table 3).

Furthermore, the cross-tabulation analysis between UB and BI revealed that 60.3% of participants with low UB exhibited low BI, while 41.2% of participants with high UB reported high BI. Similarly, 69.2% of participants with very high UB had very high BI. These findings suggest a significant positive association between UB and BI, indicating that an improvement in UB may lead to a corresponding increase in BI. (See Appendix 9 for details.)

5 Contributions and Conclusion

This study examines the intention to use AI conversational agents/chatbots in online health communities (OHCs), with a clear focus on the main constructs that influence behavior intention (BI) and use behavior (UB), as well as the moderating effects of users' intention to use these platforms. The findings of this study have significant theoretical, methodological, and managerial implications (Table 4).

The present study had several limitations that warrant consideration. These limitations encompass constraints associated with the Unified Theory of Acceptance and Use of Technology (UTAUT) conceptual model, the use of cross-sectional measurement at a single point in time, the generalizability of findings, and the inability to access low-income countries.

Given these limitations, it is advisable for future research endeavors to delve into the ethical principles and practical applications of chatbots. Additionally, these studies should explore the impact of cultural and regional factors on behavioral intention and user behavior.

In summary, our current focus on sustainability variables is undergoing expansion to incorporate various new facets. These new variables revolve around the cost-effectiveness of AI implementation in healthcare, Patient Outcomes, User experience, and long-term viability. These additions will help us gain insights into how AI technologies affect critical aspects such as diagnosis accuracy, medical error reduction, and treatment plan.

We maintain that a comprehensive investigation of variables is required to evaluate their suitability for inclusion in the proposed model, given their importance and relevance

Table 4 Theoretical, methodological, and managerial contributions

Theoretical	Methodological	Managerial
Sociomateriality concept: contributions to the theory ⊛ Understanding of users' intents with respect to deploying chatbots in OHCs Understanding how materiality affects social interactions in OHCs Sustainability concept: Contributions to the theory ⊛ contributions to the sustainability of healthcare systems by promoting efficiency, accessibility, and empowerment	UTAUT extensions: contributions to methodology: the findings revealed which factors/variables, such as PE, EE, SI, FC, FOTA, and trust, had a significant impact on the use of chatbots in OHCs	Contributions for decision-makers: integration of chatbots in OHCs in daily life of the healthcare system: this study has given healthcare professionals valuable information on the factors that influence the use of chatbots in OHCs. This study has highlighted some of the views of healthcare professionals that can be used to better understand the adoption of sustainable chatbots in OHCs in the healthcare industry
The present study made notable contribution to the UTAUT model by incorporating Trust and FOTA as the main constructs in this model, as well as additional moderators that determine users' intentions to utilize chatbots in OHCs. Moreover, the contextualization and adaptation of the UTAUT model to the healthcare industry constitutes a significant contribution that fills a critical gap in the existing literature	A significant contribution of this study was its comprehensive approach to engage members of OHCs using social media platforms. While the questionnaire was administered through Survey Monkey, all other communication with the OHCs and data analysis had to be conducted manually. The study aimed to establish a relationship with community administrators to obtain approval for message dissemination. The study actively participated in online meetings, including the Facebook community held on April 7th, 2021, which brought together individuals from OHCs around the world with an interest in the research topic. These meetings were instrumental in reaching out to the target population	The findings of the study revealed that all variables under consideration exhibited significant association with the managerial aspect of chatbot development in the European Union (EU), as evidenced by the results obtained from both Western and Eastern European nations. These contributions ultimately led to the formulation of novel policies and undertakings aimed at positioning Europe as a leading global hub for the implementation of chatbots in online help centers, whilst ensuring that AI systems are trustworthy and sustainable

in comprehending users' intentions and patterns of utilization. The incorporation of sustainability principles will remain a fundamental component of our forthcoming studies, ensuring the robustness of the model over time.

Acknowledgements This work was supported by the Erasmus+ KA2 under the project "CRISIS: Competences for Resilient Smart Cities" (Project No.: 2021-1-EL01-KA220-HED-000032257).

Appendices

Appendix 1: Presentation of UTAUT Additional Variables, Moderators, and Main Hypotheses

Variables	HYPOTHESIS
Performance expectancy	H1. PE posits a positive relationship towards the BI to use platforms/chatbots in OHCs
Effort expectancy	H2. EE posits a positive relationship towards the BI to use platforms/chatbots in OHCs
Social influence	H3. SI posits a positive relationship towards the BI to use platforms/chatbots in OHCs
Facilitating conditions	H4. FC will have a significant effect on the BI to use platforms/chatbots in OHCs
Fear of technological advancements	H5. FOTA posits a negative relationship towards the BI
Trust	H6. Trust will have a significant effect on the BI to use platforms/chatbots in OHCs
Behavioral intention	H7. BI will have a significant effect on the UB to use platforms/chatbots in OHCs
Experience in technology	H1a, H1b, H1c, H1d, H1e, and H1f: Experience in technology will moderate the relationship between PE, EE, SI, FC, FOTA and Trust respectively, to use chatbots in OHCs
Experience in chatbots	H2a, H2b, H2c, H2d, H2e, and H2f: Experience in chatbots will moderate the relationship between PE, EE, SI, FC, FOTA and Trust respectively, to use chatbots in OHCs
Geographical zones	H3a, H3b, H3c, H3d, H3e, and H3f: Geographical zones will moderate the relationship between PE, EE, SI, FC, FOTA and Trust respectively, to use chatbots in OHCs

Appendix 2: Demographic Characteristics of the Study Population

		Frequency	Percent
Gender	Female	217	49.0
	Male	204	46.0
	Prefer not to answer	22	5.0

Appendix 3: Experience Related to the Use of Chatbots in OHCs

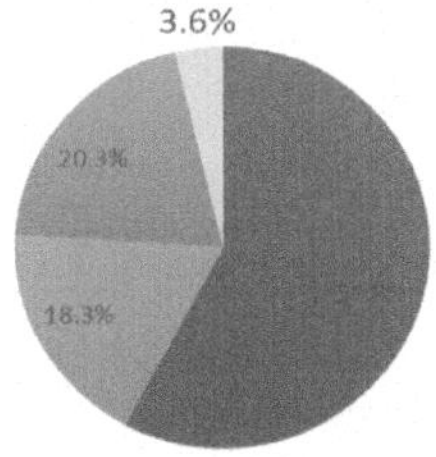

Appendix 4: Correlation Between UTAUT Scores and BI

		Behavioral Intention
Performance expectancy	Pearson correlation (r)	0.603
	P-Value	<0.001[a]
	N	443
Effort expectancy	Pearson correlation (r)	0.507
	P-Value	<0.001[a]
	N	443
Social influence	Pearson correlation (r)	0.565
	P-Value	<0.001[a]
	N	443
Facilitating conditions	Pearson correlation (r)	0.440
	P-Value	<0.001[a]
	N	443
FOTA	Pearson correlation (r)	0.084
	P-Value	0.077[a]
	N	443
Trust	Pearson correlation (r)	0.632
	P-Value	<0.001[a]
	N	443

Appendix 5: Linear Regression for the Factors Affecting BI in the European Population

Model summary				
Model	R	R Square	Adjusted R Square	Std. Error of the Estimate
1	0.781[a]	0.611	0.600	3.736[a]

a. Predictors: (Constant), trust, FOTA, effort expectancy, social influence, performance expectancy

Model		Unstandardized coefficients		Standardized coefficients	T	P-Value
		B	Std. Error	Beta		
1	(Constant)	−0.079	1.420		−0.055	0.956
	Performance Expectancy	0.381	0.051	0.431	7.408	0.000
	Effort Expectancy	0.170	0.057	0.162	3.010	0.003
	Social Influence	0.169	0.034	0.273	4.945	0.000
	FOTA	−0.121	0.058	−0.102	−2.060	0.041
	Trust	0.175	0.059	0.164	2.958	0.004

a. Dependent Variable: BI

Appendix 6: Correlation Between UTAUT Scores and Technology Use Experience

	Technology use experience	N	Mean	Std. deviation	P-Value
Effort expectancy	I have limited experience	30	23.10	7.64	<0.001[a]
	I am progressing	81	24.11	7.10	
	I consider myself an adequate user	223	25.21	6.24	
	I am an expert	109	27.63	6.14	
Facilitating conditions	I have limited experience	30	16.93	5.63	0.046[a]
	I am progressing	81	18.70	4.73	
	I consider myself an adequate user	223	18.49	4.48	
	I am an expert	109	19.48	4.35	

	Technology use experience	N	Mean	Std. deviation	P-Value
FOTA	I have limited experience	30	11.53	5.36	<0.001[a]
	I am progressing	81	10.47	4.96	
	I consider myself an adequate user	223	8.50	5.08	
	I am an expert	109	6.56	4.78	

[a]ANOVA test/significant level was set at 5% and correlations were marked in bold

Appendix 7: Correlation Between UTAUT Scores and Experience in Chatbots in OHCs

	Experience related to the use of Chatbots in OHCs	N	Mean	Std. deviation	P-Value
Performance expectancy	I have limited experience	256	20.73	7.09	<0.001[a]
	I am progressing	81	24.85	5.81	
	I consider myself an adequate user	90	21.53	7.75	
	I am an expert	16	20.13	8.85	
Effort expectancy	I have limited experience	256	24.30	6.93	<0.001[a]
	I am progressing	81	27.51	4.38	
	I consider myself an adequate user	90	26.81	6.15	
	I am an expert	16	26.13	9.24	
Social influence	I have limited experience	256	21.46	11.79	0.020[a]
	I am progressing	81	25.58	10.56	
	I consider myself an adequate user	90	24.28	10.87	
	I am an expert	16	23.94	12.97	
FOTA	I have limited experience	256	9.26	5.28	<0.001[a]
	I am progressing	81	8.89	5.47	
	I consider myself an adequate user	90	6.27	3.93	
	I am an expert	16	9.44	5.82	
Behavioral intention	I have limited experience	256	17.15	6.13	0.007[a]
	I am progressing	81	19.78	5.26	
	I consider myself an adequate user	90	18.38	6.40	
	I am an expert	16	17.19	8.27	

	Experience related to the use of Chatbots in OHCs	N	Mean	Std. deviation	P-Value
Use behavior	I have limited experience	256	10.43	3.63	0.015[a]
	I am progressing	81	11.81	3.50	
	I consider myself an adequate user	90	11.10	2.78	
	I am an expert	16	11.00	3.72	

Appendix 8: Correlation Between UTAUT Scores and Geographical Zones

	Country	N	Mean	Std. deviation	P-Value
Performance expectancy	Europe	183	20.81	6.68	<0.001[a]
	North America	105	22.87	7.10	
	Arab countries	98	20.07	7.51	
	Others	57	24.65	7.61	
Social influence	Europe	183	20.11	9.55	<0.001[a]
	North America	105	25.84	12.90	
	Arab countries	98	23.31	11.29	
	Others	57	25.56	13.15	
Facilitating conditions	Europe	183	18.08	4.00	0.041[a]
	North America	105	19.51	5.03	
	Arab countries	98	18.42	4.57	
	Others	57	19.40	5.43	
Trust	Europe	183	15.17	5.53	0.012[a]
	North America	105	16.77	6.32	
	Arab countries	98	15.38	5.79	
	Others	57	17.72	6.65	
Behavioral intention	Europe	183	17.20	5.90	0.025[a]
	North America	105	17.89	6.55	
	Arab countries	98	17.88	5.62	
	Others	57	20.07	6.96	

[a]ANOVA test/significant level was set at 5% and correlations were marked in bold

Appendix 9: Cross Tabulation Between BI and UB

		BI			Total	P-Value
		Poor	Good	Very good		
UB	Poor	249	87	77	413	<0.001[a]
		60.3%	21.1%	18.6%	100.0%	
	Good	2	7	8	17	
		11.8%	41.2%	47.1%	100.0%	
	Very good	1	3	9	13	
		7.7%	23.1%	69.2%	100.0%	
Total		252	97	94	443	
		56.9%	21.9%	21.2%	100.0%	

References

Abd-Alrazaq, A. A., Rababeh, A., Alajlani, M., Bewick, B. M., & Househ, M. (2020). Effectiveness and safety of using chatbots to improve mental health: Systematic review and meta-analysis. *Journal of Medical Internet Research, 22*(7), e16021. https://doi.org/10.2196/16021

Abedin, B., & Babar, A. (2018). Institutional vs. non-institutional use of social media during emergency response: A case of Twitter in 2014 Australian bush fire. *Information Systems Frontiers, 20*(4), 729–740. https://doi.org/10.1007/s10796-017-9787-4.

Ahad, M. A., Paiva, S., Tripathi, G., & Feroz, N. (2020). Enabling technologies and sustainable smart cities. *Sustainable Cities and Society, 61*, 102301. https://doi.org/10.1016/j.scs.2020.102301

Alahmari, N., Alswedani, S., Alzahrani, A., Katib, I., Albeshri, A., & Mehmood, R. (2022). Musawah: A data-driven AI approach and tool to co-create healthcare services with a case study on cancer disease in Saudi Arabia. *Sustainability, 14*(6), 3313. https://doi.org/10.3390/su14063313

Al-Kahtani, N., Alrawiai, S., Al-Zahrani, B. M., et al. (2022). Digital health transformation in Saudi Arabia: A cross-sectional analysis using healthcare information and management systems society's digital health indicators. *Digital Health, 8*, 20552076221117744. https://doi.org/10.1177/20552076221117742

Arfi, W. B., Nasr, I. B., Kondrateva, G., & Hikkerova, L. (2021). The role of trust in intention to use the IoT in eHealth: Application of the modified UTAUT in a consumer context. *Technological Forecasting and Social Change, 167*, 120688. https://doi.org/10.1016/j.techfore.2021.120688

Badidi, E. (2022). Edge AI and blockchain for smart sustainable cities: Promise and potential. *Sustainability, 14*(13), 7609. https://doi.org/10.3390/su14137609

Bettiga, D., & Lamberti, L. (2018). Exploring the role of anticipated emotions in product adoption and usage. *Journal of Consumer Marketing, 35*(3), 300–316. https://doi.org/10.1108/JCM-06-2016-1860

Bender, C. (2019). Artificial intelligence and the future of medicine. *Journal of the American Medical Association, 321*(23), 2275–2276. https://doi.org/10.1001/jama.2019.4914

Burton-Jones, A., Akhlaghpour, S., Ayre, S., Barde, P., Staib, A., Sullivan, C., et al. (2020). Changing the conversation on evaluating digital transformation in healthcare: Insights from an institutional analysis. *Information and Organization, 30*, 100255. https://doi.org/10.1016/j.infoandorg.2019.100255

Calvaresi, D., Calbimonte, J. P., Debusson, F., Najjar, A., & Schumacher, M. (2019). Social network chatbots for smoking cessation: Agent and multi-agent frameworks. In *IEEE/WIC/ACM International Conference on Web Intelligence* (pp. 286–292). Association for Computing Machinery, New York, NY, USA. https://doi.org/10.1145/3350546.3352532.

Cecez-Kecmanovic, D., Galliers, R. D., Henfridsson, O., Newell, S., & Vidgen, R. (2014). The sociomateriality of information systems: Current status, future directions. *MIS Quarterly, 38*(3), 809–830.

Chedrawi, C., & Haddad, G. (2022). The rise of quasi-humans in AI fueled organizations, an ultimate socio-materiality approach to the lens of Michel Serres. *Pacific Asia Journal of the Association for Information Systems, 14*(2), Article 2. https://doi.org/10.17705/1pais.14202.

Chakraborty, S., Paul, S., Ghatak, S., Pandey, S. J., Kumar, A., Singh, K. U., & Shah, M. A. (2022). An AI-based medical chatbot model for infectious disease prediction. *IEEE Access, 10*, 128469–128483. https://doi.org/10.1109/ACCESS.2022.3212852

Compeau, D. R., & Higgins, C. A. (1995). Computer self-efficacy: Development of a measure and initial test. *MIS Quarterly, 19*(2), 189–211. https://doi.org/10.2307/249688

Darnault, C., Parcollet, T., & Morchid, M. (2019). Artificial intelligence: A tale of social responsibility. *HAL*. https://hal.archives-ouvertes.fr/hal-02270410v2.

De Barcelos Silva, A., et al. (2020). Intelligent personal assistants: A systematic literature review. *Expert Systems with Applications*. https://doi.org/10.1016/j.eswa.2020.113227

Di Giacomo, G., Guerra, F., Perilli, E., & Ranieri, J. (2020). Technophobia as an emerging risk factor in aging: Investigation on computer anxiety dimension. *Health Psychology Research, 8*(1), 8207. https://doi.org/10.4081/hpr.2020.8207

Dwivedi, Y. K., Hughes, L., Ismagilova, E., Aarts, G., Coombs, C., Crick, T., et al. (2021). Artificial intelligence (AI): Multidisciplinary perspectives on emerging challenges, opportunities, and agenda for research, practice, and policy. *International Journal of Information Management*, 101994. https://doi.org/10.1016/j.ijinfomgt.2021.101994.

Faraj, S., Lakhani, K., Monteiro, E., & von Krogh, G. (2016). Online community as space for knowledge flows. *Information Systems Research, 27*(4), 668–684. https://doi.org/10.1287/isre.2016.0644

Forbes. (2022). Ripe for disruption: Artificial intelligence advances deeper into healthcare. *Forbes*. https://www.forbes.com/sites/joemckendrick/2022/11/19/ripe-for-disruption-artificial-intelligence-advances-deeper-into-healthcare/?sh=68f53bc21f5a.

Guo, X., Sun, Y., Wang, N., Peng, Z., & Yan, Z. (2013). The dark side of elderly acceptance of preventive mobile health services in China. *Electronic Markets, 23*(1), 49–61. https://doi.org/10.1007/s12525-012-0114-4

Hadley, T. D., Pettit, R. W., Malik, T., Khoei, A. A., & Salihu, H. M. (2020). Artificial intelligence in global health—A framework and strategy for adoption and sustainability. *International Journal of Maternal and Child Health and AIDS, 9*(1), 121. https://doi.org/10.21106/ijma.296.

Hesamian, M. H., Jia, W., He, X., & Kennedy, P. (2019). Deep learning techniques for medical image segmentation: Achievements and challenges. *Journal of Digital Imaging, 32*(4), 582–596. https://doi.org/10.1007/s10278-019-00227-x

Hsu, M. H., Chen, C. M., Juang, W. C., Cai, Z. X., & Kuo, T. (2021). Analyzing groups of inpatients' healthcare needs to improve service quality and sustainability. *Sustainability, 13*(21), 11909. https://doi.org/10.3390/su132111909

Jung, Y., Hur, C., & Kim, M. (2018). Sustainable situation-aware recommendation services with collective intelligence. *Sustainability, 10*(5), 1632. https://doi.org/10.3390/su10051632

Khechine, H., & Lakhal, S. (2018). Technology as a double-edged sword: From behavior prediction with UTAUT to students' outcomes considering personal characteristics. *Journal of Information Technology Education: Research, 17,* 63–102. https://doi.org/10.28945/4012.

Koulouri, T., Macredie, R. D., & Olakitan, D. (2022). Chatbots to support young adults' mental health: An exploratory study of acceptability. *ACM Transactions on Interactive Intelligent Systems, 12*(2), Article 11, 39 pages. https://doi.org/10.1145/3485874.

Leonardi, P. M. (2013). Theoretical foundations for the study of sociomateriality. *Information and Organization, 23*(2), 59–76. https://doi.org/10.1016/j.infoandorg.2013.02.002

Lee, Y., Kozar, K. A., & Larsen, K. R. (2003). The technology acceptance model: Past, present, and future. *Communications of the Association for Information Systems, 12,* 752–780. https://doi.org/10.17705/1CAIS.01250.

Maican, C. I., Cazan, A. M., Lixandroiu, R. C., & Dovleac, L. (2019). A study on academic staff personality and technology acceptance: The case of communication and collaboration applications. *Computers & Education, 128,* 113–131. https://doi.org/10.1016/j.compedu.2018.09.018

Mamoshina, P., Vieira, A., Putin, E., & Zhavoronkov, A. (2016). Applications of deep learning in biomedicine. *Molecular Pharmaceutics, 13*(5), 1445–1454. https://doi.org/10.1021/acs.molpharmaceut.5b00982

Meuter, M. L., Ostrom, A. L., Bitner, M. J., & Roundtree, R. (2003). The influence of technology anxiety on consumer use and experiences with self-service technologies. *Journal of Business Research, 56*(11), 899–906. https://doi.org/10.1016/S0148-2963(01)00276-4

Milne-Ives, M., de Cock, C., Lim, E., Shehadeh, M. H., de Pennington, N., Mole, G., et al. (2020). The effectiveness of artificial intelligence conversational agents in health care: Systematic review. *Journal of Medical Internet Research, 22*(10), e20346. https://doi.org/10.2196/20346

Muhammad, G., Hossain, M. S., & Kumar, N. (2021). EEG-based pathology detection for home health monitoring. *IEEE Journal on Selected Areas in Communications, 39*(2), 603–610. https://doi.org/10.1109/JSAC.2020.3020654

Nambisan, S., Lyytinen, K., Majchrzak, A., & Song, M. (2017). Digital innovation management: Reinventing innovation management research in a digital world. *MIS Quarterly, 41*(1), 223–238. https://doi.org/10.25300/MISQ/2017/41.1.13.

Nguyen, T., & Wang, D. (2021). An intelligent healthcare platform for predicting risk of type 2 diabetes. *Journal of Healthcare Engineering, 2021,* 1–13. https://doi.org/10.1155/2021/5529878

Niehaves, B., & Plattfaut, R. (2014). Internet adoption by the elderly: Employing IS technology acceptance theories for understanding the age-related digital divide. *European Journal of Information Systems, 23*(6), 708–726. https://doi.org/10.1057/ejis.2013.19

Osta, A., Kokkinaki, A., & Chedrawi, C. (2022). Online health communities: The impact of AI conversational agents on users. In M. Themistocleous & M. Papadaki (Eds.), *Information Systems—18th European, Mediterranean, and Middle Eastern Conference, EMCIS 2021, Proceedings* (pp. 488–501). Springer Science and Business Media Deutschland GmbH. https://doi.org/10.1007/978-3-030-95947-0_35.

Orlikowski, W. J., & Scott, S. V. (2008). Sociomateriality: Challenging the separation of technology, work and organization. *The Academy of Management Annals.* https://doi.org/10.1080/19416520802211644

Orlikowski, W. J., & Baroudi, J. J. (1991). Studying information technology in organizations: Research approaches and assumptions. *Information Systems Research, 2*(1), 1–28. https://doi.org/10.1287/isre.2.1.1

Park, E. H., Im, G., Storey, V. C., & Baskerville, R. L. (2019). Never, never together again: How postpurchase affect drives consumer outcomes within the context of online consumer support communities. *Journal of the Association for Information Systems, 20*(1), 58–104. https://doi.org/10.17705/1jais.00526.

Reinares-Lara, E., Olarte-Pascual, C., & Pelegrín-Borondo, J. (2018). Do you want to be a cyborg? The moderating effect of ethics on neural implant acceptance. *Computers in Human Behavior, 85,* 43–53. https://doi.org/10.1016/j.chb.2018.03.032

Richie, C. (2022). Environmentally sustainable development and use of artificial intelligence in health care. *Bioethics,* 36. https://doi.org/10.1111/bioe.13018.

Riffle, C. (2017). What artificial intelligence means for sustainability. *Greenbiz.* https://www.greenbiz.com/article/what-artificial-intelligence-means-sustainability.

Sillence, E. (2013). Giving and receiving peer advice in an online breast cancer support group. *Cyberpsychology, Behavior, and Social Networking, 16,* 480–485. https://doi.org/10.1089/cyber.2013.1512

Sood, S. K., Rawat, K. S., & Kumar, D. (2022). A visual review of artificial intelligence and Industry 4.0 in healthcare. *Computers and Electrical Engineering, 101,* 107948. https://doi.org/10.1016/j.compeleceng.2022.107948.

Steinert, S., & Roeser, S. (2020). Emotions, values and technology: Illuminating the blind spots. *Journal of Responsible Innovation, 7*(3), 298–319. https://doi.org/10.1080/23299460.2020.1738024

Sweeney, C., et al. (2021). Can chatbots help support a person's mental health? Perceptions and views from mental healthcare professionals and experts. *ACM Transactions on Computing for Healthcare, 2*(3), Article 25. https://doi.org/10.1145/3448871.

Thomaz, F., Salge, C., Karahanna, E., et al. (2020). Learning from the dark web: Leveraging conversational agents in the era of hyper-privacy to enhance marketing. *Journal of the Academy of Marketing Science, 48,* 43–63. https://doi.org/10.1007/s11747-019-00704-3

Tortorella, G. L., Fogliatto, F. S., Tlapa Mendoza, D., Pepper, M., & Capurro, D. (2022). Digital transformation of health services: A value stream-oriented approach. *International Journal of Production Research.* https://doi.org/10.1080/00207543.2022.2048115

Tuccillo, F., Marino, M. G., & Torri, L. (2020). Italian consumers' attitudes towards entomophagy: Influence of human factors and properties of insects and insect-based food. *Food Research International, 137,* Article 109619. https://doi.org/10.1016/j.foodres.2020.109619.

United Nations. (2015). *Transforming our world: The 2030 agenda for sustainable development.* United Nations Department of Economic and Social Affairs. https://sdgs.un.org/publications/transforming-our-world-2030-agenda-sustainable-development-17981.

Venkatesh, V., Morris, M. G., Gordon, B. D., & Davis, F. D. (2003). User acceptance of information technology: Toward a unified view. *MIS Quarterly, 27*(3), 425–478. https://doi.org/10.2307/30036540

Venkatesh, V., & Davis, F. D. (2000). A theoretical extension of the technology acceptance model: Four longitudinal field studies. *Management Science, 46*(2), 186–204. https://doi.org/10.1287/mnsc.46.2.186.11926

Viswanathan, M., & Sreekumar, A. (2019). Consumers and technology in a changing world: The perspective from subsistence marketplaces. *European Journal of Marketing, 53*(6), 1254–1274. https://doi.org/10.1108/EJM-11-2017-0826

Walsh, I. (2015). Using quantitative data in mixed-design grounded theory studies: An enhanced path to formal grounded theory in information systems. *European Journal of Information Systems,* 1-27. https://doi.org/10.1057/ejis.2014.23

Whittlestone, J., Nyrup, R., Alexandrova, A., Dihal, K., & Cave, S. (2019). *Ethical and societal implications of algorithms, data, and artificial intelligence: A roadmap for research.* Nuffield Foundation.

Winter, S. R., Rice, S., & Lamb, T. L. (2020). A prediction model of consumers' willingness to fly in autonomous air taxis. *Journal of Air Transport Management, 89,* Article 101926. https://doi.org/10.1016/j.jairtraman.2020.101926.

Yan, Z., Wang, T., Chen, Y., & Zhang, H. (2016). Knowledge sharing in online health communities: A social exchange theory perspective. *Information & Management, 53*(5), 643–653. https://doi.org/10.1016/j.im.2016.02.001

Zhang, X., Liu, S., & Chen, X. (2017). Social capital, motivations, and knowledge sharing intention in health Q&A communities. *Management Decision, 55*(7), 1536–1557. https://doi.org/10.1108/MD-10-2016-0739

Zhang, X., Liu, S., Chen, X., Wang, L., Gao, B., & Zhu, Q. (2018). Health information privacy concerns, antecedents, and information disclosure intention in online health communities. *Information & Management, 55*(4), 482–493. https://doi.org/10.1016/j.im.2017.11.003

Zhu, Y., Wang, R., & Pu, C. (2022). "I am chatbot, your virtual mental health adviser": What drives citizens' satisfaction and continuance intention toward mental health chatbots during the COVID-19 pandemic? An empirical study in China. *Digital Health, 8.* https://doi.org/10.1177/20552076221090031.

Sustaining Innovation Performance in Service Organizations: The Role of Total Quality Management Practices and Strategic Alignment

Maryam Al Kamzari and Farzana Asad Mir

Abstract

The turbulent business environment of today poses research questions about our understanding of the processes at play regarding achieving competitive advantage through sustainable innovation performance. "Total quality management (TQM)" is known to provide improvement in all business processes including innovation, but how do these interact with the requirements of an uncertain environment is unclear. We examine the role of an organization's strategic alignment in the relationship between TQM practices and innovation performance. Four TQM practices were selected from the "Malcolm Baldrige National Quality Award" model; "customer focus, human resource management, continuous improvement and leadership" and their relationship was tested with strategic alignment and innovation performance. Analysis of data from 357 respondents working in service organizations in the Sultanate of Oman, showed that TQM practices have a significant and positive impact on strategic alignment, and strategic alignment partially mediates the relationship between TQM practices and innovation performance. This study uncovers interesting findings about the influence of TQM practices on organizational strategic alignment and how that in turn impacts the product, process, and administrative innovation performance. The findings inform managers looking for enhanced innovation performance in the current uncertain business environment about the important role of TQM practices.

M. Al Kamzari (✉) · F. A. Mir
The British University in Dubai, Dubai, United Arab Emirates
e-mail: 20190861@student.buid.ac.ae

F. A. Mir
e-mail: farzana.mir@buid.ac.ae

Keywords

Total quality management · Strategic alignment · Innovation performance

1 Introduction

Managers are increasingly becoming aware of the key role of strategic alignment in enhancing firm performance, providing impetus for increased research in this area (Kearns & Lederer, 2003). Studies show that strategic alignment supports organizations in acquisition and improvement of assets and abilities, strengthening their market position (Rondinelli et al., 2001). "Total Quality Management (TQM)" focuses on individuals' mobilization, resource consistency in the firm, and management leadership encompassing the whole organization, in order to provide customer satisfaction (Belén Escrig-Tena, 2004). While there is substantial research on both strategic alignment and innovation performance individually, limited research directly examines how strategic alignment influences innovation performance. Moreover, the connection between TQM and strategic alignment is an area with limited research. While TQM is widely recognized for its positive impact on various aspects of organizational performance, there is a scarcity of studies that thoroughly explore how TQM practices influence an organization's strategic alignment. This represents a research gap that requires more attention.

From an innovation management perspective, TQM has the potential to influence managerial practices in innovative organizations. In other words, using TQM as a lens to see innovation management will help understand how to enhance and support innovation from standpoint of quality management practice in an organization (Qiang, 2020). Recent studies emphasize the need for more comprehensive research that investigates the direct and indirect

M. Amer et al. (eds.), *Territorial Landscapes Solutions for Sustainable Energy and Food Security*,
Sustainable Landscape Planning and Natural Resources Management, https://doi.org/10.1007/978-3-031-82419-7_13

mechanisms through which TQM practices influence innovation (Pekovic & Galia, 2009). These studies suggest that while TQM is known to enhance organizational performance, the exact ways in which it fosters innovation are not fully understood. Therefore, this study aims to examine the relationship between TQM and strategic alignment, and the relationships of these constructs with innovation performance. We further examine the mediation effect of strategic alignment between TQM and innovation performance.

2 Literature Review

2.1 Total Quality Management (TQM)

Waldman (1994) stated that TQM is an integrated organizational strategy aimed at enhancing the quality of services and products. TQM is defined as an approach that enhances the management of the organization leading to improvement in quality of organizational outcomes/processes and increase in satisfaction of its customers as well as profits (Kivimäki et al., 1997). Firms implement TQM for increasing and enhancing its performance by raising the competitive advantage and becoming innovative (Abdullah et al., 2009). Studies show that organizations should follow TQM as a management strategy (Garvin, 1988). "Malcolm Baldrige National Quality Award (MBNQA)" is frequently used as a TQM framework which was developed by "The National Institute of Standards and Technology" (Kent et al., 2006). The award criteria were essentially established for measuring the QM application level in organizations (Bon & Mustafa, 2013).

There are several TQM practices that aid the organizations to achieve customer satisfaction and continuous improvement, such as, provision of top management support facilitating achievement of organizations' goals, creation of a culture that supports employees' involvement in decision-making and development of effective stakeholder relationships. Hansson and Klefsjö (2003) identified the following six values of TQM based on TQM core values perspective: Top management commitment, customers focus, everybody's commitment, process focus, continuous improvement, and facts based decision. Prior studies show that the most significant TQM dimensions are:" (a) top management support, (b) continuous improvement, (c) employee involvement, and (d) customer focus" (Juran, 1988a).

Based on the American MBNQA, the current study used four dimensions of TQM, namely customer focus (CF), human resource management (HRM), continuous improvement (CNI), and leadership (LDR). These dimensions have also used in some studies, such as Abbas (2020) and Hung et al. (2011).

The principle of CF encourages innovation by introducing and developing new customer expectations and needs, by providing new products in order to adapt to the change of the market needs continuously (Juran, 1988b). In organizations, the CF is an important factor leading to better business outcomes (Omar et al., 2018). An enhanced focus and coordination with customers leads to increasing and improving their satisfaction (Wan et al., 2021).

In TQM organizational capabilities and resources are managed in an efficient manner (Shafiq et al., 2019). Another TQM factor is HRM that concentrates on the development of human resources through active participation and training of employees in operational issues (Abbas, 2020). It focuses on employee performance recognition and empowerment, quality awareness and responsibility, and reward mechanism. Moreover, HRM comprises a variety of practices that enhance organizational development such as recruitment procedure, efficient training, involvement, safety and health practices, empowerment, teamwork, recognition, etc. In addition, it is widely acknowledged that well-trained, committed, and satisfied human resources improve organizational performance (Singh et al., 2018).

The CNI concept is mainly intended at streamlining or simplifying a process making it efficient (Lawler III, 1994). Prajogo and Sohal (2003) stated that firms might continually follow faulty processes and because of that continuous improvement is essential for innovation and performance. The basic pillar of TQM philosophy is that employees should be encouraged by their firms to continuously search for new improvements and ideas. Hence, focusing on continuous improvement as a practice of TQM, firms can ensure quality in their innovation processes and outcomes (Singh et al., 2018).

Successful TQM implementation relies heavily on LDR. The following skills are essential for the executives: quality leadership knowledge, prioritize ability, systems knowledge, quality knowledge, and change management knowledge. TQM implementations are considered more successful when top managers expose all these skills (Antonaros, 2010). Organization LDR is an important TQM factor as shown in many empirical TQM studies (Xiong et al., 2017).

2.2 Innovation Performance

The term innovation encompasses generation of new ideas, knowledge applications, methods, and skills which can influence the competitiveness of the organization and produce unique competencies (Andersson et al., 2008). Focusing on products and services as outputs of organizations the literature categorizes as product and process innovation. Product innovation means providing different services or products, while process innovation focuses on methods or processes adopted for producing services or products (Propris, 2002). Valle and Vázquez-Bustelo (2009) define product innovation as the development of new products (or services) by integrating new knowledge with existing technology of now in use products. Whereas, process innovation is defined by Gatignon et al. (2002) as "the new or significantly improved elements" of service operations or an organization's production to decrease costs and/or increase the quality of product/service. Administrative innovation is how the new ideas are generated and applied to improve systems, processes and structures of the organization itself (Weerawardena, 2003). This study applies these three innovation types to investigate the role of strategic alignment in enacting the relationship between TQM and innovation performance.

2.3 TQM and Innovation Performance

There is an evidence of a positive relationship between TQM and innovation performance (Shan et al., 2016). There are some TQM practices that were observed to be critical factors to innovation success such as focus on quality, customer, process and employees, as well as continuous improvement and leadership (Gustafson & Hundt, 1995). Moreover, TQM practices significantly impact product innovation with obvious differentiation of innovation speed, whereas TQM contributes to quick time to market (Flynn, 1994). In other words, organizations that implement TQM will enable the sharing of knowledge and encourage a rich situation for innovation (Kim et al., 2012). By defining the aim of innovation, i.e., the desire to increase efficiency or performance, organization leaders can develop innovation strategies to be implemented (Shin et al., 2022). Mustafa and Bon (2013) have argued that there is a relationship between TQM and service and product innovation (radical innovation, incremental innovation, and administrative innovation).

2.4 Strategic Alignment

This concept is defined by Venkatraman and Camillus (1984) as "a fit between a firm's strategy" with its environment both externally and internally. The shaping of strategic adaption requires alignment of the organization's resources with external market threats and opportunities (Dickson et al., 2006). Internal alignment includes the organizational culture and structure, which enables the firm to situate itself relative to market opportunities and forms the intrafirm competitiveness (Ku et al., 2011). Given that the market is in constant turmoil, organizations must be keen to address these challenges quickly through innovation, digital transformation, new skills and commitment to continue their success and presence in the market (Scuotto et al., 2022). When it is hard to adjust or align quickly to external changes, changes in environments become challenging (Atuahene-Gima & Li, 2004).

Prior studies have found that strategic alignment is dependent upon a few basic constraints (Tallon, 2011). To start with, there are still a few inquiries about how and why value is created by strategic alignment. From the stockholders' view, the value is not limited to financial outcomes rather it encompasses the non-financial outcomes as well (Amit & Zott, 2001). Practically all the current explorations have analyzed the effect of alignment on financial outcomes, whereas the value is recommended as a superior indicator that does not restrict itself to financial outcomes only (Acosta et al., 2011).

3 Conceptual Framework

3.1 Theory

According to contingency theory "organizational effectiveness results from fitting characteristics of the organization to contingencies that reflect the situation of the organization" (Donaldson, 2001). Based on this conceptualization the strategy is a contingent factor, not a universal concept (Donaldson, 2006), and a fit is required with internal and external context (Acur et al., 2012) to achieve desired performance. The contingency theory contends that there is no perfect fit for all organizations. The theory further highlights that strategic alignment isn't all-inclusive and is related to the context in which it is applied to and hence it is a necessary requirement for achieving competitive advantage as it provides a fit between organization resources and capabilities with the environment in which it resides (Rashidirad et al., 2013).

Using this theoretical foundation, a conceptual research framework is developed and then empirically tested. The proposed conceptual framework is shown in Fig. 1, whereas the hypotheses are described below.

3.2 TQM and Strategic Alignment

Misalignment of strategy may lead to the lack of agility (Alpkan et al., 2007) and loss of competitiveness and absence of market and customer focus (Bagnoli & Giachetti, 2015). To improve strategic alignment, resource-limited organizations use "off-the-shelf" TQM approaches

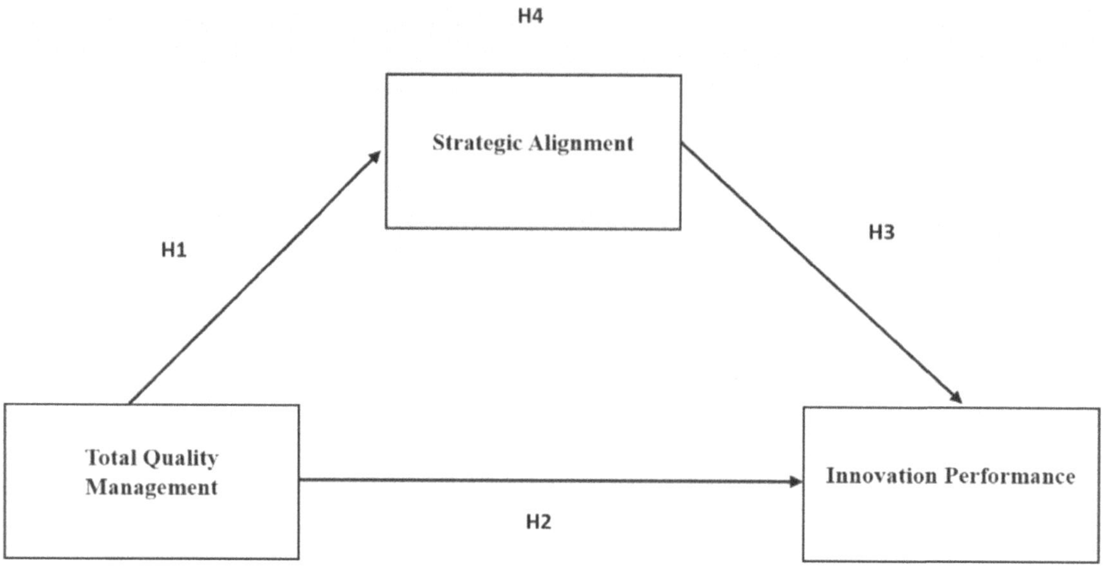

Fig. 1 Research framework

which is counter-intuitive and do not support strategic alignment requirement (McAdam et al., 2019).

In order to achieve better firm performance a fit is required between strategy and the changes in the market dynamics. This can only occur by maintaining an awareness and alignment with change occurrence which consists of both "continuous change and discontinuous change" (Din & Cheema, 2013). Continuous change is harnessed with TQM and discontinuous change begins with innovation. At the point when an organization needs to make some sort of progress agreeing to the external environment forces or any other forces, it needs to adopt both types of change by embracing TQM and innovation, in order to find alignment with external forces of change.

Differentiation on innovation, and marketing, as well as cost leadership as strategic dimensions would help the firm to develop strategic alignment (Strandholm et al., 2004). Therefore, we propose:

H1: TQM has a positive impact on strategic alignment.

3.3 TQM and Innovation Performance

TQM places significant emphasis on customer satisfaction, human resource management, employee engagement as well as continuous process improvement. Prior literature regarding the relationship between innovation performance and TQM shows a positive relationship between these constructs, i.e., the companies implementing TQM create an environment conducive to innovation because TQM aligns with innovative principles. By focusing on understanding customer needs, organizations are prompted to identify new customer expectations and requirements (Din & Cheema, 2013). Hence, we propose:

H2: TQM has a positive impact on innovation performance.

3.4 Strategic Alignment and Innovation Performance

Competitive advantage is driven by innovation and organizational flexibility (Zahra & George, 2002). Strategic flexibility is a dynamic capability enabling an organization to effective utilization of assets and reconfiguration of processes and procedures (Eisenhardt & Martin, 2000). Innovations are complex in nature and normally occur within an organization which is adaptable and flexible. Firms with greater flexibility can rapidly adjust their item advancement choices to respond to the changing market needs for their potential benefit (Grewal & Tansuhaj, 2001). There is evidence of positive influence of strategic alignment on "new product development" (Kandemir & Acur, 2012) and innovation and its capabilities (Fan et al., 2013). Discontinuous organizational changes begin with innovation and they are infrequent and internal. It is normally an unexpected and unusual change. The discontinuous change is reflected in an organization's ability to align with the fast changes in the market and coming up with replacement strategies and outcomes (Din & Cheema, 2013).The third hypothesis of this study is:

H3: Strategic alignment has a positive impact on innovation performance.

Based on the first three hypotheses, we propose the following hypothesis:

H4: Strategic alignment mediates the relationship between TQM and innovation performance.

4 Research Method and Results

4.1 Data

Prior to distributing the questionnaire to the target population, a pretest was conducted. The pretest involved a review by two managers who meticulously checked the questionnaire for any issues, such as unclear or ambiguous questions, response options, or instructions. The questionnaire was then translated into Arabic by a proficient translator, as it was intended for distribution in the Sultanate of Oman, where the Arabic language is the native language.

This research utilized a cross-sectional quantitative approach, collecting data through an online questionnaire. Non-probability purposive sampling was employed. The study focused on participants at the organizational level, including Chief Executive Officers (CEOs), General Director, Director, and Head of Department. Email addresses of organizations were obtained from the official web sites and their official social media accounts. The survey link was also shared with the target respondents through the LinkedIn platform. Out of 500 questionnaires that were distributed, 357 completed responses were received, which were used for further analysis. Table 1 summarizes the sample characteristics.

Table 1 Socio-demographic sample structure ($n = 357$)

Socio-demographic variables	Category	N	Percentage (%)
Job sector	Government sector	283	79.3
	Private sector	74	20.7
Job position	CEO	21	5.9
	General director	70	19.6
	Director	160	44.8
	Head of department	106	29.7
Experience	"Less than 5 years"	37	10.4
	"5—less than 10 years"	96	26.9
	"10—less than 15 years"	96	26.9
	"15—less than 20 years"	63	17.6
	"20 years and above"	65	18.2

4.2 The Measurement Instrument

The measurement scales used to evaluate the model were taken from the existing literature and were adapted to the context under study.

4.3 Analysis of Data

Analysis was done using SPSS 26 and SMART-PLS 3.3.7. To examine "common method bias", we did the "Harman's single-factor test" (Greene & Organ, 1973). This analysis is a post hoc procedure to check whether a single factor emerges. The first component variance is equal to 49% which is less than 50% of the variance (Podsakoff & Organ, 1986) excluding any CMV issues. Table 2 shows a comprehensive representation of how the constructs were operationalized.

4.4 Measurement Model

In this study, all first-order and second- order constructs were reflective in nature. A disjoint two-stage approach was employed for analysis. In the first stage, the measurement model was assessed for the first-order components. In the second stage, the scores for the first-order latent variables (LDR, CF, HRM, CN, PRCINV, PRDINV, and ADMINV) were derived from the preceding step, utilizing the first-order indicator scores.

The measurement model provides reliability and validity information. Validity assesses the scale's ability to capture the concept accurately, while reliability demonstrates the consistency and stability of the scale (Sekaran, 2005).

For reflective latent constructs, reliability at both the construct and item levels was measured using "composite reliability" and "Cronbach's Alpha". The values were more than the accepted threshold of 0.70 (Hair et al., 2014). "Indicator reliability" was also evaluated, whereby item loadings should exceed 0.7, following Hair et al. (2011). Nearly all items displayed factor loadings greater than 0.7, except for three items, CF2, HRM2, and ADMINV4, which had loadings of 0.683 and 0.626, and 0.670 respectively as shown in Table 3. These three items were not eliminated from the analysis, as their removal did not lead to an improvement in reliability.

For "convergent validity", the "Average Variance Extracted (AVE)" values were examined which should be >0.5 (Fornell & Larcker, 1981). As shown in Table 4, all the constructs met this criterion.

Table 2 Operationalisation of main constructs

Construct	Operationalization	Code for constructs	Code for indicators
Total quality management	Second-order reflective	TQM	LDR, CF, HRM, CNI
Leadership	First-order reflective	LDR	LDR1, LDR2, LDR3, LDR4, LDR5
Customer focus	First-order reflective	CF	CF1, CF2, CF3, CF4, CF5, CF6, CF7
Human resource management	First-order reflective	HRM	HRM1, HRM2, HRM3, HRM4, HRM5, HMR6, HMR7, HRM8
Continuous improvement	First-order reflective	CNI	CNI1, CNI2, CNI3, CNI4
Innovation performance	Second-order reflective	INVPR	PRCINV, PRDINV, ADMINV
Process innovation	First-order reflective	PRCINV	PRCINV1, PRCINV2, PRCINV3, PRCINV4
Product innovation	First-order reflective	PRDINV	PRDINV1, PRDINV2, PRDINV3, PRDINV4, PRDINV5, PRDINV6
Administrative innovation	First-order reflective	ADMINV	ADMINV1, ADMINV2, ADMINV3, ADMINV4
Strategic alignment	First-order reflective	STA	STA1,STA2, STA3, STA4

To assess "discriminant validity", several methods were employed. First, indicators' outer loading were found to be greater than their cross-loadings (Farrell, 2010). Second, the square root of the AVE exceeded the inter-construct correlations (Fornell & Larcker, 1981). Finally, the "Heterotrait-Monotrait ratio (HTMT)" values were below 0.9, which indicates sufficient "discriminant validity" (Hair et al., 2014). The results, are shown in Tables 4 and 5.

Likewise, in the validation of the second-order reflective latent constructs (Fig. 2), which are TQM and INVPR, "composite reliability" and "Cronbach's Alpha" values were >0.70 (Hair et al., 2014). The item loadings also passed the 0.7 threshold, following the criteria laid out by Hair et al. (2011). Furthermore, the AVE value was >0.5 (Refer to Table 6). Additionally, the results confirmed that there was "discriminant validity" (See Table 7).

4.5 Structural Model

By using "Partial Least Squares Structural Equation Modeling (PLS-SEM)", the structural model was examined for several important factors. These factors include the "coefficient of determination (R^2)", "predictive relevance (Stone-Geisser Q^2)", and the effect size. The structural model is illustrated in Fig. 3, displaying the "standardized path coefficients (β)" and the "variance explained in the endogenous variables" (R^2). To ensure the robustness of our findings, we performed a bootstrap analysis with 5000 resamples, which helped us extract the significance of our estimates, indicated by the t-statistics.

Figure 3 and Table 8 indicate the summary of the hypothesis results. The results indicate that TQM significantly impact strategic alignment and innovation performance ($\beta = 0.734$, $p < 0.001$; $\beta = 0.614$, $p < 0.05$), providing support for H1 and H2. The results also show that strategic alignment significantly impact innovation performance (H3: $\beta = 0.283$; $p < 0.001$). Moreover, the effect of TQM on innovation performance was found to be partially mediated by strategic alignment (H4: $\beta = 0.285$, $t = 6.240$, $p < 0.05$).

As Fig. 3 illustrates, TQM explains about 54% of the variance in strategic alignment, and both TQM and strategic alignment explain about 71% of the variance in innovation performance showing a moderate predictive power of the model (Chin, 1998).

Furthermore, we assessed the Q^2 values, which serve as indicators of the accuracy achieved by the model's predictions. A Q^2 value >0 confirms predictive relevance (Henseler et al., 2009) The Q^2 values for strategic alignment and innovation performance were 0.399 and 0.5981, respectively, showing satisfactory results.

Additionally, for model fit "Standardized Root Mean Square Residual (SRMR)" test was done. For a good fit SRMR should be ≤ to 0.08 (Hu & Bentler, 1999). In this study, the SRMR was calculated as 0.046 showing that the model aligned well with the observed data, strengthening the validity of our analysis.

Table 3 Factor loadings of first order constructs items

Constructs	Items	Loadings
Leadership (LDR)	LDR1	0.764
	LDR2	0.837
	LDR3	0.773
	LDR4	0.832
	LDR5	0.825
Customer focus (CF)	CF1	0.793
	CF2	0.683
	CF3	0.784
	CF4	0.831
	CF5	0.816
	CF6	0.813
	CF7	0.798
Human resource management (HRM)	HRM1	0.763
	HRM2	0.626
	HRM3	0.801
	HRM4	0.821
	HRM5	0.827
	HRM6	0.726
	HRM7	0.823
	HRM8	0.768
Continuous improvement (CNI)	CNI1	0.806
	CNI2	0.85
	CNI3	0.882
	CNI4	0.808
Strategic alignment (STA)	STA1	0.853
	STA2	0.9
	STA3	0.835
	STA4	0.873
Process innovation (PRCINV)	PRCINV1	0.791
	PRCINV2	0.847
	PRCINV3	0.862
	PRCINV4	0.845
Product innovation (PRDINV)	PRDINV1	0.878
	PRDINV2	0.895
	PRDINV3	0.912
	PRDINV4	0.843
	PRDINV5	0.867
	PRDINV6	0.883
Administrative innovation (ADMINV)	ADMINV1	0.814
	ADMINV2	0.86
	ADMINV3	0.885
	ADMINV4	0.67

5 Discussion

5.1 Theoretical Implications

This study explains whether the firm's strategic alignment influences innovation performance and whether or how TQM influences strategic alignment. According to Baldwin and Johnson (1996) the TQM adoption as a management strategy could provide the necessary differentiation between organizations having higher innovation levels compared to the organizations that lag behind in innovation. Contingency theory provided a valid theoretical foundation to explain that strategic alignment has a partial mediator role in the relationship between TQM and innovation performance. It shows how strategic alignment achieves value for innovation performance. The results of the theoretical model provide support for strategic alignment in the sense that it is showing the effect on the three different types of innovation, process, product, and administrative. Moreover, this study uncovers interesting findings on the impacts of TQM on strategic alignment.

5.2 Practical Implications

The study results highlight the benefits of TQM as it improves strategic alignment and positively affects innovation performance. These results show that organizations should commit to implementing TQM in order to improve their strategic alignment level as well as innovation performance. As several quality models are available for implementing TQM, a good starting point for organizations would be to conduct gap analysis based on an appropriate framework and then work on the identified deficiencies while building on to the existing strengths.

The study also identified and found support for the mediating role of strategic alignment in the relationship between TQM and innovation performance. This indicates that a successful TQM implementation would enhance strategic alignment, which in turn has a positive impact on innovation performance. Hence, the findings provide confidence to the firms' managers to concentrate on the effective implementation of TQM if they are looking to enhance innovation performance.

5.3 Recommendations and Future Research

Organizations should institutionalize TQM to enhance innovation performance. Embedding TQM practices

Table 4 Assessment of reliability, convergent and discriminant validity of reflective constructs

Constructs	1	2	3	4	5	6	7	8
(1) Leadership	**0.807**							
(2) Customer focus	0.777	**0.79**						
(3)Human resource management	0.77	0.766	**0.772**					
(4)Continues improvement	0.665	0.723	0.728	**0.837**				
(5) Strategic alignment	0.622	0.677	0.663	0.668	**0.866**			
(6) Process innovation	0.704	0.723	0.763	0.72	0.657	**0.837**		
(7) Product innovation	0.64	0.646	0.67	0.637	0.692	0.804	**0.88**	
(8) Administrative innovation	0.659	0.619	0.719	0.638	0.683	0.752	0.772	**0.812**
Mean	3.44	3.57	3.14	3.78	3.57	3.24	3.11	3.08
AVE	0.651	0.624	0.596	0.701	0.749	0.700	0.774	0.659
"Cronbach's alpha"	0.866	0.899	0.902	0.857	0.888	0.857	0.942	0.826
"Composite reliability"	0.903	0.920	0.921	0.903	0.923	0.903	0.954	0.884

Note Root square of AVE in bold

Table 5 Discriminant validity of first-order reflective constructs, heterotrait-monotrait ratio (HTMT)

Constructs	1	2	3	4	5	6	7	8
(1) Leadership								
(2) Customer focus	0.876							
(3) Human resource management	0.871	0.846						
(4) Continues improvement	0.772	0.823	0.829					
(5) Strategic alignment	0.708	0.755	0.74	0.765				
(6) Process innovation	0.817	0.822	0.867	0.84	0.752			
(7) Product innovation	0.708	0.697	0.724	0.708	0.754	0.895		
(8) Administrative innovation	0.761	0.697	0.813	0.736	0.786	0.877	0.865	

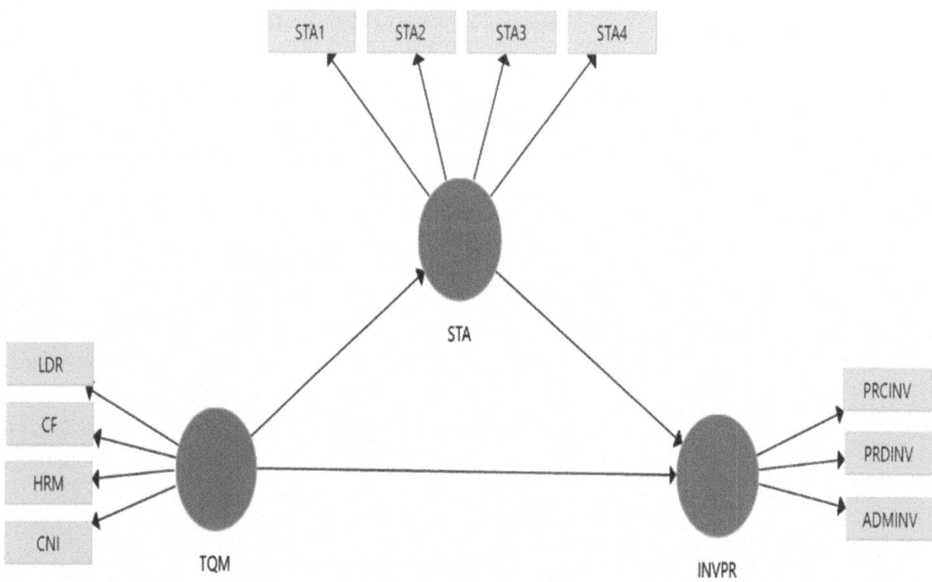

Fig. 2 Second stage of disjoint two-stage approach

deeply within operations, guided by comprehensive quality models like "Malcolm Baldrige National Quality Award (MBNQA)", can yield significant benefits. Strategic alignment is crucial as it mediates the relationship between TQM and innovation. Regular reviews and adjustments to maintain this alignment are essential. Fostering a culture of continuous improvement can drive efficiency and effectiveness. Encouraging cross-functional collaboration ensures that TQM initiatives are integrated across the organization, leading to better outcomes. Investing in training programs

Table 6 Psychometric properties of the second-order reflective constructs

Second-order construct	Items	Loadings	"Cronbach's alpha"	"Composite reliability"	AVE
TQM	LDR	0.894	0.919	0.943	0.804
	CF	0.911			
	HRM	0.912			
	CNI	0.869			
INVPR	PRCINV	0.927	0.912	0.945	0.851
	PRDINV	0.929			
	ADMINV	0.910			

Table 7 Discriminant validity of second-order reflective construct, HTMT

Constructs	TQM	STA	INVPR
TQM			
STA	0.815		
INVPR	0.896	0.811	

equips employees with the skills and knowledge necessary for successful TQM implementation, enhancing overall quality and innovation performance. Future research should include both service and product organizations and explore TQM's application in various departments for a comprehensive understanding.

Regular monitoring and evaluation of TQM initiatives with established metrics and benchmarks are essential to ensure their effectiveness. Continuous feedback and improvement cycles can help refine these initiatives to achieve desired outcomes. These recommendations aim to help organizations enhance their strategic alignment and innovation performance through effective TQM practices.

Moreover, future studies should test the role of strategic alignment in the relationship between TQM financial/nonfinancial performance. Several factors driving the

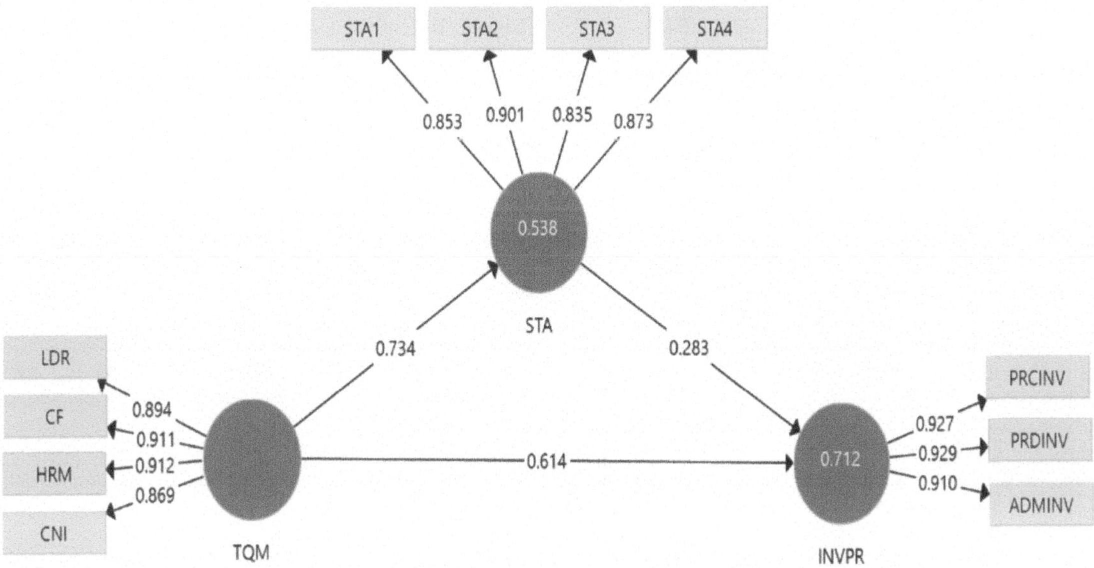

Fig. 3 Results of research model

Table 8 Hypothesis test results

Hypothesis	Coefficient	p-value	Support for hypothesis
H1: TQM ⟶strategic alignment	0.734	$P<0.001$	Supported
H2: TQM ⟶ innovation performance	0.614	$P<0.001$	Supported
H3: Strategic alignment ⟶ innovation performance	0.283	$P<0.001$	Supported
H4: Strategic alignment mediates the relationship between TQM and innovation performance	0.208	$P<0.001$	Supported

application and effectiveness of TQM from both internal and external perspectives can be explored. For example, how TQM is executed in different industry areas with various degrees of dynamism and uncertainty. Additionally it would be useful to investigate the application of TQM in various departments or functions within the same organization, for instance, among production and marketing departments.

6 Conclusion

In conclusion, our results have verified the proposed conceptual model that hypothesized that TQM has a direct effect on strategic alignment and innovation performance. Moreover, strategic alignment has a partial mediator role in the relationship between TQM and innovation performance. The results provided support for the importance of strategic alignment by showing its effect on the three different types of innovation.

References

Abbas, J. (2020). Impact of total quality management on corporate green performance through the mediating role of corporate social responsibility. *Journal of Cleaner Production, 242*, 118458.

Abdullah, M., Uli, J., & Tari, J. (2009). The relationship of performance with soft factors and quality improvement. *Total Quality Management, 20*, 735–748.

Acosta, P. S., Colomo-Palacios, R., & Loukis, E. N. (2011). A review of the RBV of the firm within the e-business literature: What's next. *Interdisciplinary Journal of Research in Business, 1*, 45–52.

Acur, N., Kandemir, D., & Boer, H. (2012). Strategic alignment and new product development: Drivers and performance effects. *Journal of Product Innovation Management, 29*, 304–318.

Alpkan, L., Yilmaz, C., & Kaya, N. (2007). Market orientation and planning flexibility in SMEs: Performance implications and an empirical investigation. *International Small Business Journal, 25*, 152–172.

Amit, R., & Zott, C. (2001). Value creation in e-business. *Strategic Management Journal, 22*, 493–520.

Andersson, M., Lindgren, R., & Henfridsson, O. (2008). Architectural knowledge in inter-organizational IT innovation. *The Journal of Strategic Information Systems, 17*, 19–38.

Antonaros, M. E. (2010). *Gender differences in leadership style: A study of leader effectiveness in higher education*, ProQuest Information and Learning.

Atuahene-GIMA, K., & Li, H. (2004). Strategic decision comprehensiveness and new product development outcomes in new technology ventures. *Academy of Management Journal, 47*, 583–597.

Bagnoli, C., & Giachetti, C. (2015). Aligning knowledge strategy and competitive strategy in small firms. *Journal of Business Economics and Management, 16*, 571–598.

Baldwin, J. R., & Johnson, J. (1996). Business strategies in more-and less-innovative firms in Canada. *Research Policy, 25*, 785–804.

Belénescrigtena, A. (2004). TQM as a competitive factor: A theoretical and empirical analysis. *International Journal of Quality & Reliability Management, 21*(6), 612–637.

Bon, A. T., & Mustafa, E. M. (2013). Impact of total quality management on innovation in service organizations: Literature review and new conceptual framework. *Procedia Engineering, 53*, 516–529.

Chin, W. W. (1998). The partial least squares approach to structural equation modeling. *Modern Methods for Business Research, 295*(2), 295–336.

Dickson, D. R., Ford, R. C., & Upchurch, R. (2006). A case study in hotel organizational alignment. *International Journal of Hospitality Management, 25*, 463–477.

Din, M. S., & Cheema, K. U. (2013). Strategic change: A study of TQM and innovation, pp. 254–260.

Donaldson, L. (2001). *The contingency theory of organizations*. Sage London UK.

Donaldson, L. (2006). The contingency theory of organizational design: challenges and opportunities. In *Organization design: The evolving state-of-the-art* (pp. 19–40). Springer.

Eisenhardt, K. M., & Martin, J. A. (2000). Dynamic capabilities: What are they? *Strategic Management Journal, 21*, 1105–1121.

Fan, Z., Wu, D., & Wu, X. (2013). Proactive and reactive strategic flexibility in coping with environmental change in innovation. *Asian Journal of Technology Innovation, 21*, 187–201.

Farrell, A. M. (2010). Insufficient discriminant validity: A comment on Bove, Pervan, Beatty, and Shiu (2009). *Journal of Business Research, 63*(3), 324–327.

Flynn, B. B. (1994). The relationship between quality management practices, infrastructure and fast product innovation. *Benchmarking for Quality Management and Technology, 1*(1), 48–64.

Fornell, C., & Larcker, D. F. (1981). Evaluating structural equation models with unobservable variables and measurement error. *Journal of Marketing Research, 18*, 39–50.

Garvin, D. A. (1988). *Managing quality: The strategic and competitive edge*. Simon and Schuster.

Gatignon, H., Tushman, M. L., Smith, W., & Anderson, P. (2002). A structural approach to assessing innovation: Construct development of innovation locus, type, and characteristics. *Management Science, 48*, 1103–1122.

Greene, C. N., & Organ, D. W. (1973). An evaluation of causal models linking the received role with job satisfaction. *Administrative Science Quarterly*, 95–103.

Grewal, R., & Tansuhaj, P. (2001). Building organizational capabilities for managing economic crisis: The role of market orientation and strategic flexibility. *Journal of Marketing, 65*, 67–80.

Gustafson, D. H., & Hundt, A. S. (1995). Findings of innovation research applied to quality management principles for health care for. *Health Care Management Review, 20*, 16–33.

Hair, J. F., Ringle, C. M., & Sarstedt, M. (2011). PLS-SEM: Indeed a silver bullet. *Journal of Marketing Theory and Practice, 19*(2), 139–152.

Hair, J. F., Jr., Sarstedt, M., Hopkins, L., & Kuppelwieser, V. G. (2014). Partial least squares structural equation modeling (PLS-SEM): An emerging tool in business research. *European Business Review, 26*(2), 106–121.

Hansson, J., & Klefsjo, B. (2003). A core value model for implementing total quality management in small organisations. *The TQM Magazine, 15*(2), 71–81.

Henseler, J., Ringle, C. M., & Sinkovics, R. R. (2009). The use of partial least squares path modeling in international marketing. In *New challenges to international marketing* (Vol. 20, pp. 277–319). Emerald Group Publishing Limited.

Hu, L. T., & Bentler, P. M. (1999). Cutoff criteria for fit indexes in covariance structure analysis: Conventional criteria versus new alternatives. *Structural Equation Modeling: A Multidisciplinary Journal, 6*(1), 1–55.

Hung, R. Y., Lien, B., Yang, B., Wu, C., & Kuo, Y. (2011). Impact of TQM and organizational learning on innovation performance in the high-tech industry. *International Business Review, 20*, 213–225.

Juran, J. M. (1988a). *Juran on planningfor quality*. Juran Institute Inc.
Juran, J. M. (1988b). *Juran on planning for quality*, Collier Macmillan.
Kandemir, D., & Acur, N. (2012). Examining proactive strategic decision-making flexibility in new product development. *Journal of Product Innovation Management, 29*, 608–622.
Kearns, G. S., & Lederer, A. L. (2003). A resource-based view of strategic IT alignment: How knowledge sharing creates competitive advantage. *Decision Sciences, 34*, 1–29.
Kent, K., Chevalier, S., & Grance, T. (2006). Guide to integrating forensic techniques into incident. *Guide to Integrating Forensic Techniques into Incident Response*, 800–886.
Kim, D., Kumar, V., & Kumar, U. (2012). Relationship between quality management practices and innovation. *Journal of Operations Management, 30*, 295–315.
Kivimäki, M., Mäki, E., Lindström, K., Alanko, A., Seitsonen, S., & Järvinen, K. (1997). Does the implementation of total quality management (TQM) change the wellbeing and work-related attitudes of health care personnel? *Journal of Organizational Change Management, 10*(6), 456–470.
Ku, E. C., Wu, W., & Lin, A. (2011). Strategic alignment leverage between hotels and companies: The buyer–supplier relationship perspective. *International Journal of Hospitality Management, 30*, 735–745.
Lawler, E. E., III. (1994). Total quality management and employee involvement: Are they compatible? *Academy of Management Perspectives, 8*, 68–76.
McAdam, R., Miller, K., & Mcsorley, C. (2019). Towards a contingency theory perspective of quality management in enabling strategic alignment. *International Journal of Production Economics, 207*, 195–209.
Mustafa, E., & Bon, A. (2013). *The relationship between TQM and innovation in service organizations: Quantitative findings*.
Omar, S., Ariffin, Z., Abdullah, A., Mahadi, N., Mokhtar, F., & Azman, N. (2018). Conceptualising the influence of total quality management and organizational performance in Klang Valley service industry. *Advanced Science Letters, 24*, 3240–3243.
Pekovic, S., & Galia, F. (2009). From quality to innovation: Evidence from two French employer surveys. *Technovation, 29*(12), 829–842.
Podsakoff, P., & Organ, D. (1986). Self-reports in organizational research: Problems and prospects. *Journal of Management, 12*, 531–544.
Prajogo, D., & Sohal, A. (2003). The relationship between TQM practices, quality performance, and innovation performance. *International Journal of Quality and Reliability Management, 20*(8), 901–918.
Propris, L. (2002). Types of innovation and inter-firm co-operation. *Entrepreneurship and Regional Development, 14*, 337–353.
Qiang, F. (2020). Building innovation management assessment theory framework through the lens of TQM. *International Journal of Business Research, 20*, 31.
Rashidirad, M., Soltani, E., & Syed, J. (2013). Strategic alignment between competitive strategy and dynamic capability: Conceptual framework and hypothesis development. *Strategic Change, 22*, 213–224.
Rondinelli, D., Rosen, B., & Drori, I. (2001). The struggle for strategic alignment in multinational corporations: Managing readjustment during global expansion. *European Management Journal, 19*, 404–416.
Sekaran, U. (2005). *Research methods for business: A skill building approach* (4th ed.). Wiley.
Scuotto, V., Magni, D., Palladino, R., Nicotra, M., & Change, S. (2022). Triggering disruptive technology absorptive capacity by CIOs. Explorative research on a micro-foundation lens. *Technological Forecasting and Social Change,* 174, 121234.
Shafiq, M., Lasrado, F., & Hafeez, K. (2019). The effect of TQM on organisational performance: Empirical evidence from the textile sector of a developing country using SEM. *Total Quality Management and Business Excellence, 30*, 31–52.
Shan, A., Ahmad, M., & Nor, N. (2016). The mediating effect of innovation between total quality management (TQM) and business performance. In *IOP Conference Series: Materials Science and Engineering* (Vol. 160, No. 1, p. 01201). IOP Publishing.
Shin, J., Kim, Y., Jung, S., & Kim, C. (2022). Product and service innovation: Comparison between performance and efficiency. *Journal of Innovation & Knowledge, 7*(3), 100191.
Singh, V., Kumar, A., & Singh, T. (2018). Impact of TQM on organisational performance: The case of Indian manufacturing and service industry. *Operations Research Perspectives, 5*, 199–217.
Strandholm, K., Kumar, K., & Subramanian, R. (2004). Examining the interrelationships among perceived environmental change, strategic response, managerial characteristics, and organizational performance. *Journal of Business Research, 57*, 58–68.
Tallon, P. (2011). Value chain linkages and the spillover effects of strategic information technology alignment: A process-level view. *Journal of Management Information Systems, 28*, 9–44.
Valle, S., & Vázquez-Bustelo, D. (2009). Concurrent engineering performance: Incremental versus radical innovation. *International Journal of Production Economics, 119*, 136–148.
Venkatraman, N., & Camillus, J. (1984). Exploring the concept of "fit" in strategic management. *Academy of Management Review, 9*, 513–525.
Waldman, D. (1994). The contributions of total quality management to a theory of work performance. *Academy of Management Review, 19*, 510–536.
Wan, X., Liu, X., Du, Z., & Du, Y. (2021). A novel model used for assessing supply chain sustainability integrating the ANP and ER approaches and its application in marine ranching. *Journal of Cleaner Production, 279*, 123500.
Weerawardena, J. (2003). The role of marketing capability in innovation-based competitive strategy. *Journal of Strategic Marketing, 11*, 15–35.
Xiong, J., Deng, Y., Zhang, M., & Zhang, Z. (2017). Quality management practices and their effects on the performance of public hospitals. *International Journal of Quality and Service Sciences, 9*(3/4), 383–401.
Zahra, S., & George, G. (2002). Absorptive capacity: A review, reconceptualization, and extension. *Academy of Management Review, 27*, 185–203.

The manufacturer's authorised representative in the EU is Springer
Nature Customer Service Centre GmbH, Europaplatz 3, 69115 Heidelberg,
Germany. If you have any concerns regarding our products, please
contact ProductSafety@springernature.com

Printed and bound by CPI Group (UK) Ltd, Croydon, CR0 4YY
05/10/2026
02228167-0001